W9-ALV-170

Pro | ENGINEER®
W I L D F I R E™

Tutorial
and MultiMedia CD

Text by
Roger Toogood, Ph.D., P. Eng.

Multimedia CD by
Jack Zecher

ISBN: 1-58503-113-5

SDC
PUBLICATIONS

Schroff Development Corporation

www.schroff.com
www.schroff-europe.com

Preface

This tutorial was created to introduce new users to Pro/ENGINEER® and has been updated for Wildfire. This release incorporates major revisions to the software, particularly in regards to the user interface, which continues to become more Windows-like and easier to use. The tutorial covers the major concepts and frequently used commands required to advance from a novice to an intermediate user level. Major topics include part and assembly creation, and creation of engineering drawings. The major functions that make Pro/E a parametric solid modeler are illustrated.

Although the commands are presented in a click-by-click manner, an effort has been made, in addition to showing/illustrating the command usage, to explain *why* certain commands are being used and the relation of feature selection and construction to the overall part design philosophy. Moreover, since error recovery is an important skill, considerable time is spent exploring the created models (in fact, intentionally inducing some errors), so that users will become comfortable with the "debugging" phase of model creation. In my experience of teaching numerical methods courses, debugging and error recovery are sadly neglected when students are first taught a programming course, and the same probably applies to CAD.

This series of lessons was originally written for students in the Engineering Graphics and Design course (MecE 265) offered in the Mechanical Engineering program at the University of Alberta **<http://www.mece.ualberta.ca/courses/mec265/>**. This is a required course taken by all students entering Mechanical Engineering, and is the only engineering graphics course in their program. We have been using Pro/ENGINEER since the fall of 1996. Students enter the course with a broad range of backgrounds - some have previous CAD experience, while others have only an introductory programming course. Since students taking the course have a wide range of abilities both in spatial visualization and computer skills, the approach taken here is meant to allow accessability to persons of all levels. These lessons, therefore, were written for new users with no previous experience with CAD, although some familiarity with computers is assumed.

This book is **NOT** a reference for Pro/ENGINEER. Coverage of all that Pro/E has to offer within a single (even not-so-thin) volume is quite impossible. Since Release 18 of Pro/E, all the several thousand pages of reference manuals and official Pro/E documentation are available on-line. Access to this on-line help has been considerably improved in Wildfire, which offers good search tools and cross-referencing to allow users to find relevant material quickly and in much more detail than can be presented here.

The lessons in this tutorial are meant to be covered sequentially. Discussion of commands is, for the most part, restricted to their use within the context of the lesson (a Just-in-Time delivery!). For this reason, many options to commands are not dealt with in detail all in the same place in the text, as is done in the on-line reference material. Such a discussion would interrupt the flow of the work. Although the index provides locations within the text where a command is used, this is not exhaustive, but rather meant to act as a quick reference or reminder.

Since these tutorials were first written (for Release 16), quite a number of changes have occurred. Major changes in the previous editions have primarily involved the reorganization of material.

With the current Wildfire release, much new material was required to deal with the new interface and its operation. A number of minor errors have been corrected and some additional comments have been inserted at various places in the text to clarify the discussion.

In the previous tutorials (for Releases 2000i^2 and 2001), major changes occurred in the first couple of lessons to deal with the rapidly evolving user interface in Pro/E. Minor effects were felt throughout the lessons. These changes revolve around usability issues - the same basic functionality of Pro/E is still there. New tools for dealing with parent/child relations, a new Sketcher interface, part and drawing templates, new pull-down menus, a new hole creation dialog window, drag-and-drop reordering in the model tree, many new pop-up menus available with a right mouse click, and so on, were discussed. All in all, the number of command selections and mouse clicks required to use Pro/E was reduced dramatically with the new interface. Another new addition was the availability of a Web site with VRML models of the Panavise project parts and assembly; see <**http://www.sdcpublications.com/tutorial**>.

The multimedia CD-ROM produced by Jack Zecher at IUPU-Indianapolis is included with the goal of providing students with multi-modal learning tools and experiences. The CD should help significantly in getting "up the learning curve." The CD follows the text very closely, and although it does not go into all the detail contained in the text, provides an excellent overview of the material in each lesson. We expect that many students will find it advantageous to go through the CD presentation for each lesson (or part thereof) prior to working through the lesson in detail.

In the previous (2001) version of the Tutorial, there were a couple of major changes in organization of the material. A completely new lesson (#1) was prepared for introducing new users to the program with the aim of giving them a better overview of program operations (primarily view and display management) and capability. Hopefully, this will help users keep aware of the big picture even when they are deeply involved in the detailed inner workings later. The lesson on sweeps and blends (important features, but perhaps a luxury in an introductory series) was moved to the end of the tutorial. Several sections in other lessons were modified significantly. For example, some tutorial activities involving difficult sketches were either removed or reworked. Some more attention was paid to the Intent Manager in Sketcher. A new section on creating radial patterns using make datums was added. In the hard copy edition of the book an Appendix was added that deals with interface customization.

Notes for the **WILDFIRE** Edition

The organization of the lessons remains the same as the previous edition. The major changes in the presented material result from the new Wildfire interface - the dashboard, extensive use of toolbars, the right mouse button pop-up menus, Navigator and Browser windows, new dialog windows, and so on. The concept of "Direct Modeling" is introduced and used in several places. Sketcher is now utilized exclusively with Intent Manager turned on (the default). Some new guidelines are presented for more logical and efficient use of Intent Manager.

Note to Instructors

The Engineering Graphics and Design course at the U of A is a one-term course of 12 weeks, with two lectures and a three hour lab every week. Most lecture time is dedicated to introducing students to the theory and practice of creating engineering drawings (reading drawings, visualization of shape from multiview drawings, layout of multiview drawings, detailing and sectioning practice and standards, and so on). Half the weekly lab activities are devoted to free hand drawing activities such as sketching pictorial views or freehand layout of multiview drawings and solving drawing problems (eg. missing view problems). The other half of the lab each week is spent working with Pro/E, primarily going through these lessons. Students must spend an additional 2 to 3 hours per week with these tutorials.

The tutorials consist of the following:
    1 lesson introducing the program and its operation
    6 lessons on features used in part creation
    1 lesson on modeling utilities
    1 lesson on creating engineering drawings
    2 lessons on creating assemblies and assembly drawings
Each of these will take between 2 to 4 hours to complete (thus usually requiring some time spent out of the regularly scheduled computer lab time). The time required will vary depending on the ability and background of the student. Moreover, additional time would be beneficial for experimentation and exploration of the program. Most of the material can be done by the student on their own time, however there are a few "tricky" bits in some of the lessons. Therefore, it is important to have teaching assistants available (preferably right in the computer lab) who can answer special questions and especially bail out students who get into trouble. Most common causes of confusion are due to not completing (or even doing!) the lessons or digesting the material. This is not surprising given the volume of new information or the lack of time in students' schedules. However, I have found that most student questions are answered within the lessons. In addition to the tutorials, some class time (two to three hours) over the duration of the course will be invaluable in demonstration and discussion of some of the broader issues of feature-based modeling. It takes a while for students to realize that just creating the geometry is not sufficient for a design model, and the notion of design intent needs careful treatment and discussion.

It is important for students to keep up the pace with the Pro/E lessons through the course. To that end, laboratory exercises have involved short quizzes (students produce written answers to questions chosen at random from the end of each lesson), creating models of parts sketched on the whiteboard in isometric or multiview, or brought into the lab (usually large models made of styrofoam). Of these, the latter two activities seemed to have been the most successful. It appears that many students, after having gone through the week's lesson (usually only once, and very quickly) do not absorb very much. The second pass through the lesson usually results in considerably more retention. Students really don't feel comfortable or confident until they can make parts from scratch on their own. Each lesson concludes with a number of simple "exercise" parts that can be created using new commands taught in that lesson. In addition to these, a project is also included that consists of a number of parts that are introduced with the early lessons and finally assembled at the end. The Panavise remains my favorite choice for this project, since its parts have a wide variety of feature types and degree of difficulty. Note that VRML models are available on the Web (address given above) for visualization of the Panavise parts and assembly. It would be most beneficial, however, if students could have at their disposal a physical model (of the Panavise or some other object) which they can "reverse engineer".

As a last note, at the U of A the EGD course is a prerequisite to the first "design" course. That course involves a team design-and-build project in which one of the deliverables is a Pro/E model, complete with detailed drawings. We are fortunate, as well, to have access to a rapid prototyping machine on which students can build parts for their project, using models created in Pro/E. Besides being an excellent way to reinforce the learning started in the EGD course, this type of follow-up also gives the students an opportunity to really use Pro/E as it was meant to be - as a design tool.

Acknowledgments

The inspiration for these lessons was based on the Web pages produced by Jessica LoPresti, Cliff Phipps, and Eric Wiebe of the Graphic Communications Program, Department of Mathematics, Science and Technology Education at North Carolina State University. Permission to download and modify their pages is gratefully acknowledged. Since that time (July, 1996) the tutorials have been rewritten/updated seven times: initially to accommodate our local conditions and then for Releases 18, 20, 2000i (with another foray into PT/Modeler in between), 2000i^2, 2001, and now Wildfire. All of the figures are new and discussion of the commands is considerably amplified.

Some of the objects and parts used in these tutorials are based on illustrations and problem exercises in **Technical Graphics Communication** (Irwin, 1995) by Bertoline, Wiebe, *et al.* This book is an excellent source for examples and additional exercises in part and assembly modeling, and drawing creation.

The Panavise project in this tutorial is based on a product patented by Panavise Products, Inc. and is used with the express written permission of Panavise Products, Inc., Reno, Nevada. The name Panavise is a registered trademark of Panavise Products, Inc., Reno, Nevada, and is used with the express written permission of Panavise Products, Inc. Such permission is gratefully acknowledged.

Any other similarity of objects, parts, and/or drawings in this tutorial is purely coincidental and unintentional.

These tutorials (for Release 16) were first written as Web pages and released in September, 1996. In the 16 months they were available on the Web, they received over 30,000 hits from around the world. This number is indeed gratifying in itself, but in addition, a number of users (students, instructors, industrial users, even a patent lawyer!) have returned comments on the lessons, which are gratefully acknowledged. Constructive comments and suggestions continue to arrive regularly from readers all around the world. Many thanks to all these people, who are too numerous to list here.

I would like to thank Ian Buttar, Alan Wilson, and Pierre Genest for their help in the operation of the computer lab and to the students of the Engineering Graphics and Design course for their comments and suggestions. Notwithstanding their assistance, any errors in the text or command sequences are those of the author!

Acknowledgment is due to Stephen Schroff at SDC for his continued efforts in taking this work to a wider audience.

Jack Zecher at IUPUI has, once again, done a great job producing the accompanying CD-ROM.

I greatly appreciated a visit to PTC in June, 2002, to get a preview of Wildfire. I would like to thank Ken Page for arranging this, and all the folks at PTC who made the trip worthwhile. Also, thanks are due to Kris Pyle (Base3 Product Development Group) for making sure my account is always up-to-date.

Finally, Elaine, Kate, and Jenny continue their record of unwavering support and understanding for my many hours away from the family while working on this project. They continue to tolerate my terrible jokes (although the groans are getting louder)! Now that the basement renovations are finished, I can work mostly at home. Thank you, Jenny, for the tea and cookies. Also, thanks are due again to our good friends, Jayne and Rowan Scott, for their continued support and enthusiasm.

To users of this material, I hope you enjoy the lessons.

RWT
Edmonton, Alberta
8 April 2003

Jack Zecher writes:

I would like to express my thanks to Rob Wolter for his continued excellent work on the narration portion of the CD-ROM.

A very special note of appreciation to my wife Karen, for her understanding and tolerance of my schedule during the preparation of this CD-ROM.

Jack Zecher
IUPUI
May, 2003

About the CD-ROM

The CD-ROM was developed on a PC compatible machine running under the Windows XP operating system. In addition, it has been tested and found to run properly on Windows NT, Windows 98 and Windows 95 machines. On most machines the program will self start when you place the CD in the CD-ROM drive. If it does not, you should use "Windows Explorer" or "My Computer" to open the CD-ROM device, and then double click on the executable file *ProWild.exe*.

The recommended minimum hardware requirements are as follows:
Pentium 133 Mhz
4X CD-ROM drive
32 megabytes of RAM
1024 x 768 video resolution with 16 bit color
Sound card

You may need to make the following adjustments in order to improve the performance of the CD-ROM player under Windows 95 and Windows 98. From the **Start Menu** select the following options:

Settings > Control Panel > System > Performance
File System... > CD-ROM

then set the ***Supplemental cache size:*** to "maximum", and the ***Optimize access pattern for:*** to "No read ahead".

In order to control the volume, you will have to manually adjust the volume control on your speakers, or click on the speaker icon, usually located in the lower right hand corner of the task bar. If you are using headphones, make sure to plug them into the sound card and not the headphone jack located on the CD-ROM drive. Also, if you are using headphones, be aware that excessive sound volumes for even short periods of time can be dangerous - check the volume level before you put the headphones on.

TABLE OF CONTENTS

INTRODUCTION to Pro/ENGINEER

Lesson 1 : User Interface, View Controls and Model Structure

Lesson 2 : Creating a Simple Object (Part I)

Lesson 3 : Creating a Simple Object (Part II)

Lesson 6 : Datums and Sketcher Tools

Lesson 7 : Patterns and Copies

Lesson 8 : Engineering Drawings

Lesson 9 : Assembly Fundamentals

Lesson 10 : Assembly Operations

Lesson 11 : Sweeps and Blends

Appendix : Interface Customization

Index

This page left blank.

INTRODUCTION to
Pro I ENGINEER®
W I L D F I R E ™

A Few Words Before You Dive In...

This tutorial was written for new users who are getting started with Pro/ENGINEER Wildfire (PTC, Needham, MA <**www.ptc.com**>). The lessons in this book will introduce you to the basic functionality of the program and are meant to be used alongside the running Pro/E software. Description of command sequences is accompanied by a discussion of where the commands fit into an overall modeling strategy. In addition to learning *what* each command or function does, it is important to understand *why* it is used. We will sometimes make intentional errors so that we can discover how Pro/E responds. **Therefore, just clicking through the command sequences given here is not enough; you will learn the material best if you take time along the way to read the text carefully and think about what you are doing and observing how Pro/E responds**. You will also learn considerably more by exploring the program on your own and experimenting with the commands and options.

You are about to learn how to use one of the most sophisticated and powerful solid modeling programs available. It may be the most complex software you will ever use. It's power derives from its extremely rich command set, that understandably requires quite a long time to master. Pro/E's learning curve has the reputation of being very steep although this will diminish with the arrival of this exciting new release. The goal of this tutorial is to help you with this learning as effectively and efficiently as possible. Learning Pro/E is a challenging task, but not impossible. Do not be discouraged, as you will find it well worth the effort.

Please note that this is not a reference manual. Not all the available commands in Pro/E are covered (by a long shot!), nor will a comprehensive discussion of the myriad available options be attempted. The tutorial is meant only to get you started and to give you a solid foundation on which to build further knowledge. Nonetheless, upon completion of these lessons (in about 30 or 40 hours from now), you should:

1. be able to create models of relatively complex parts and assemblies.
2. know how to produce the related detailed engineering drawings.
3. understand the terminology used in Pro/E.
4. understand the design philosophy and methods embedded in Pro/E.

The last two are important so that you can understand the on-line reference documentation and explore other commands and functions in Pro/E.

In the early lessons and as each new function is introduced, commands are presented in considerable detail to explain what is going on and why. As you progress through the lessons, you will be given fewer details about commands that have been covered previously. For example, in Lesson #2 we find out how to create the default datum planes, mouse click by mouse click. Later on, you will be asked to "Create the default datum planes" assuming that you know how to do that. Thus, the tutorials build off each other and are meant to be done in the order presented. It is important for you to go through the lessons in sequence and to have a good understanding of the material before you go on to the next lesson.

You may have to go through each lesson (or some portions) more than once to gain an acceptable level of understanding. Each lesson has some questions and exercises at the end to allow you to check your knowledge of the concepts and commands and to give you a starting point for your own exploration of the program. No answers are given here for these questions - you will learn the material best if you have to dig them out for yourself! In doing so, you will undoubtedly uncover additional commands and options. This sort of "on-the-fly" discovery is a never-ending activity even with experienced Pro/E users because it is such a big program. Finally, each lesson concludes with a project activity that will result in the creation of a simple assembly.

The images presented here should correspond with those obtained in the Pro/E windows, and can be used to check your work as you proceed through the lessons. Figures in this document, however, are only available in black-and-white, whereas color plays an important role in the Pro/E screen. In addition to distinguishing between different parts and making "pretty pictures", color also is used to indicate the meaning of lines (edge, axis, datum curve, hidden line, and so on). Where a line interpretation may be ambiguous in the black and white version here, the figures are labeled with the appropriate line color or have been modified to show different thickness. Also, some modifications have been made to the default system font in order to make the figures clearer.

These lessons were developed using the Windows NT™ version of the software, however operation under Unix should be practically identical. The version used was production build 2002490.

What *IS* Pro/ENGINEER?

Actually, Pro/E is a suite of programs that are used in the design, analysis, and manufacturing of a virtually unlimited range of products. Its field of application is generally mechanical design, although recent additions to the program are targeted at ship building as well[1]. In these lessons, we will be dealing only with the major front-end module used for part and assembly design and model creation, and production of engineering drawings. There are a wide range of additional modules available to handle tasks ranging from sheet metal operations, piping layout, mold design, wiring harness design, NC machining, and other functions. Sensitivity studies and design optimization based purely on geometry are handled by a model called Behavioral Modeling Extension (BMX). Mechanism design, kinematics, and animation is accomplished using the Mechanism Design Extension (MDX). An add-on package, Pro/MECHANICA (also

---

[1] People who work in architectural or civil engineering design would most likely not use Pro/E, as its design and functions do not lend themselves directly to those activities.

from Parametric Technology)[2], integrates with Pro/E to perform structural analysis (static stress and deformation, buckling and fatigue analysis, vibration), thermal analysis, and dynamic motion analysis of mechanisms. Pro/MECHANICA can also do sensitivity studies and design optimization, based on the model created in Pro/E.

In a nutshell, Pro/ENGINEER is a *parametric, feature-based solid modeling* system.

"Feature-based" means that you create your parts and assemblies by defining high level and physically meaningful features like extrusions, sweeps, cuts, holes, slots, rounds, and so on, instead of specifying low-level geometry like lines, arcs, and circles. This means that you, the designer, can think of your computer model at a very high level, and leave all the low-level geometric detail for Pro/E to figure out. Features are specified by setting values and attributes of elements such as reference planes or surfaces, direction of creation, pattern parameters, shape, dimensions, and others. Features can either add or subtract material from the model, or be simple non-solid geometric entities like references axes and planes.

"Parametric" means that the physical shape of the part or assembly is driven by the values assigned to the attributes (primarily dimensions) of its features. You may define or modify a feature's dimensions or other attributes at any time (within limits!). Any changes will automatically propagate through your model. You can also relate the attributes of one feature to another. For example, if your design intent is such that a hole be centered on a block, you can relate the dimensional location of the hole to the block dimensions using a numeric formula; if the block dimensions change, the centered hole position will be re-computed automatically.

"Solid Modeling" means that the computer model you create is able to contain all the "information" that a real solid object would have. It has volume and therefore, if you provide a value for the density of the material, it has mass and inertia. Unlike a surface model, if you make a hole or cut in a solid model, a new surface is automatically created and the model "knows" which side of this surface is solid material. The most useful thing about solid modeling is that it is impossible to create a computer model that is ambiguous or physically non-realizable, such as the "object" shown in the figure at the right. This figure shows what appears to be a three-pronged tuning fork at the left end, but only has two square prongs coming off the handle at the right end. With solid modeling, you cannot

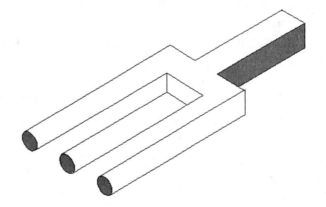

The 3-Pronged Blivot -
A Non-realizable Object

---

[2] A companion book, *The Pro/MECHANICA Tutorial* is also available from the publisher, Schroff Development Corp.

create a "model" such as this that could not physically exist. This is quite easy to do with just 2D, wireframe, or even surface modeling.

Whether or not the part could actually be manufactured is another story. Here is a cut-away view of a physically possible part, but don't take this to the machine shop and ask them to machine the cavity inside the part! Pro/E will let you make this model, but concerns of manufacturability are up to you.

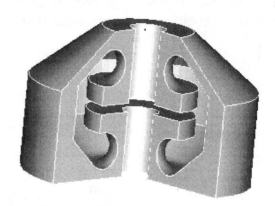

Could your machine shop make this?

An important aspect of feature-based modeling in Pro/E is the concept of **parent/child relationships**. Without going in to a lot of detail at this time, a child feature is one that references a previously created (parent) feature. For example, the surface of a block might be used as a reference plane to create a slot. A change to the parent feature will potentially affect the child. For example, deleting a parent feature will delete all its children since one or more references required to create the children would no longer exist. Pro/E has special functions available to manage parent/child relationships. This can get pretty complicated with a complex model (a good reason to try to keep your models simple!), so we will leave the details for later. However, you should keep parent/child relations in mind when you are specifying feature references for a new feature you are creating: If the parent feature is temporary or is likely to change, what effect will this have on the children? Will the references still correctly capture your design intent?

Once your model is created, it is very easy to get Pro/E to produce fully detailed standard format **engineering drawings** almost completely automatically. In this regard, Pro/E also has **bidirectional associativity** - this means you can change a dimension on the drawing and the shape of the model will automatically change, and vice versa. To a new user of the program, this is almost magic!

Of course, few parts live out their existence in isolation. Thus, a major design function accomplished with Pro/E is the construction of assemblies of parts. Assembly is accomplished by specifying physically-based geometric constraints (insert, mate, align, and so on) between part features. Of course, drawings of assemblies can also be created. With assemblies you can see how the different parts will fit together or interfere with each other, or see how they move with respect to each other, for example, in a linkage assembly.

With Wildfire, Pro/E has implemented more wide-ranging and significant changes in the user interface. The program is more Windows-like than ever. A number of new tools and software features have been introduced which make the program easier to use. A new mode of operation (called "Direct Modeling") has been introduced. At the same time, for power users, there are a large number of shortcuts (think "right mouse button") which can speed up your work quite a lot.

These have made the program easier to use (the interface style will be quite familiar to Windows users) and added a lot of visual excitement to working with the program. Another key aspect of the program is its readiness for collaboration of users over the internet. We will not be delving into these tools here, but instead concentrate on stand-alone usage to create models and assemblies.

If you do not at some point say (or at least think) "WOW!" while learning how to use Pro/E, then you are very hard to impress indeed.

This sounds like it's pretty complicated!...

It is important to realize that you won't be able to master Pro/E overnight, or even after completing these lessons. Its power derives from its flexibility and rich set of commands. It is natural to feel overwhelmed at first! With not too much practice, however, you will soon become comfortable with the basic operation of the program. As you proceed through the lessons, you will begin to get a feel for the operation of the program, and the philosophy behind feature based design. Before you know it, you'll feel like a veteran and will gain a tremendous amount of personal satisfaction from being able to competently use Pro/E to assist you in your design tasks. Some work done by students after completing this tutorial is featured in a Project Gallery, available on the Web at the URL
<**http://www.mece.ualberta.ca/courses/mec265/vrprojects.htm**>.

You will find that using Pro/E is quite different from previous generation CAD programs. This is a case where not having previous CAD experience might even be an asset since you won't have to unlearn anything! For example, because it is a solid modeling program, all your work is done directly on a 3D model rather than on 2D views of the model. Spatial visualization is very important and, fortunately, the Pro/E display is very easy to manipulate. Secondly, as with computer programming, with Pro/E you must do a considerable amount of thinking and planning ahead (some fast free-hand sketching ability will come in handy here!) in order to create a clean model of a part or assembly. Don't worry about these issues yet - they will not interfere with your learning the basic operation of the program. As you become more adept with Pro/E, you will naturally want to create more complex models. It is at this time that these high-level issues will assert themselves. In the meantime, have fun and practice, practice, practice.

Overview of the Lessons

A brief synopsis of the lessons in this tutorial is given below. Each lesson should take at least 2 to 3 hours to complete - if you go through the lessons too quickly or thoughtlessly, you may not understand or remember the material. For best results, it is suggested that you scan/browse ahead through each lesson completely before going through it in detail. The enclosed CD-ROM[3], which gives a multimedia overview of each lesson, has been created for just this task.

---

[3] The CD-ROM by Jack Zecher at IUPUI is available only with the hard copy version of this tutorial. For further information, contact the publisher at <**www.schroff.com**>.

You will then have a sense of where the lesson is going, and not be tempted to just follow the commands blindly. You need to have a sense of the whole forest when examining each individual tree!

In order to complete the lessons, you will need to install some tutorial files on your hard disk. These are also included on the enclosed CD-ROM, or can be downloaded from the Web. Further instructions for this are in Lesson #1.

On the pages following are some brief snapshots of the lessons:

Lesson 1 - User Interface, View Controls and Model Structure

How to start Pro/E; representation of Pro/E command syntax; command flow in Pro/E; special mouse functions; Pro/E windows; controls for managing the view and display of objects; the model tree; how parts and assemblies are structured.

Lesson 2 - Creating a Simple Object (Part I)

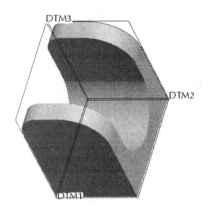

Creating a simple part using sketched features; datum curves; Sketcher and Intent Manager are introduced; sketching constraints, alignments, and procedures; feature database functions are introduced; part templates.

Lesson 3 - Creating a Simple Object (Part II)

Placed features (hole, chamfer, round) are added to the block created in Lesson #2; listing and naming features; modifying dimensions; adding relations to control part geometry; more Sketcher tools; implementing *design intent*.

Lesson 4 - More Features for Creating Parts

A new part is modeled using a number of different feature creation commands and options: both sides protrusions, an axisymmetric (revolved) protrusion, a cut, rounds, and chamfer. More Sketcher tools. Edge sets. Mirrored features. Model analysis tools. We will intentionally make some modeling errors to see how Pro/E responds.

Lesson 5 - Modeling Utilities, Parent/Child Relations, and the 3 R's

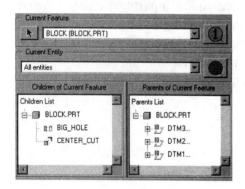

These utilities are used to investigate and edit your model: changing references, change feature shapes, changing the order of feature regeneration, changing feature attributes, and so on. Suppressing and resuming features. If your model becomes even moderately complex, you will need to know how to do this!

Lesson 6 - Sketcher Tools and Datum Planes

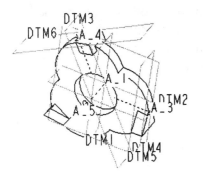

More tools in Sketcher are introduced, including sketching relations. The mysteries of datum planes and make datums are revealed! What are they, how are they created? How are they used to implement design intent?

Lesson 7 - Patterns and Copies

Creating a counterbored hole and hole notes. Patterns (one-dimensional or two-dimensional); radial patterns of placed and sketched features. Pattern groups. Copies using translation, rotation, or mirroring.

Lesson 8 - Creating an Engineering Drawing

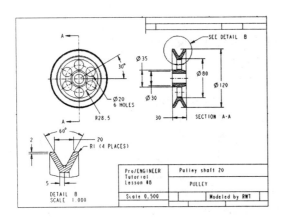

This lesson will introduce you to the process of making dimensioned engineering drawings. Two new parts are created (both parts will also be used in Lesson #9 on assemblies). Much of the work in creating the drawing is done by Pro/E, although a fair amount of manual labor must go into improving the cosmetics of the drawings.

Lesson 9 - Assembly Fundamentals

This lesson will show you how to create an assembly from previously created parts. This involves creating placement constraints that specify how the parts are to fit together. Assigning appearances (colors).

Lesson 10 - Assembly Operations

This lesson will show you how to make modifications to the assembly created in Lesson #9. This includes changing part dimensions, adding assembly features, suppressing and resuming components, creating exploded views, and creating an assembly drawing. Display styles.

Lesson 11 - Sweeps and Blends

These are the most complicated (ie. flexible and powerful) features covered in these lessons. They are both types of solid protrusions, but can also be used to create cuts and slots.

Once again, as you go through these lessons, take the time to explore the options available and experiment with the commands. You will learn the material best when you try to apply it on your own ("flying solo"), perhaps trying to create some of the parts shown in the exercises at the end of each lesson.

On-Line Help

Should you require additional information on any command or function, Pro/E comes with extensive Web-based on-line help. This contains the complete text of *all* reference manuals for the software. There are several ways you can access the on-line help. These are presented in Lesson #1.

To those of you who have read this far: Congratulations! You are probably anxious to get going with Pro/E. Let's get started...

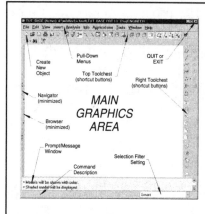

Lesson 1

User Interface, View Controls and Model Structure

Synopsis

Starting Pro/E; command syntax; mouse functions; view and display controls; model structure of parts and assemblies; the model tree; obtaining hard copy; on-line help.

Overview of this Lesson

We are going to cover a lot of introductory ground in this lesson with two main objectives. Our first is to introduce you to the user interface and locate the commands for controlling the display of parts and assemblies. You will need to be quite comfortable with these controls so that later on we can concentrate on commands for actually creating new objects or dealing with other issues of running the program.

Our second objective is to explore how models of parts and assemblies are structured[1]: what features are used, how are they ordered, and so on. It is important to keep in mind that creating a Pro/E model involves much more than simply producing geometry or creating pretty pictures. A model can have a number of purposes and end users: engineering analysis and visualization, production of drawings, manufacturing and production planning, marketing, and so on. If the model is centrally involved in design of a new product (which is typically a very iterative process), then we must ensure that it is simple, flexible, and robust to the inevitable modifications that will occur as the design evolves. Plus, there are often several different ways to create the desired geometry. Obviously, if you only know one way of doing something, your options will be limited! If you know several, which one should you use, and why? Furthermore, since a great deal of design these days is done in groups and teams, it is inevitable that you will be passing your models on to someone else. It must be easy for them to figure out how your model was made. *Simple .. flexible .. robust*. These goals are not that much different from those of writing a computer program. In fact, you might think of model creation as "programming" the

---

[1] An old saying, attributed to Lao Tzu, goes something like: "Give a man a fish and you feed him for a day. Teach him how to fish and you feed him for a lifetime." Perhaps the best way to get started is just to be shown the fish!

geometry engine to produce the product you want. As with computer programming, you will be a much more effective Pro/E user if you take some time to plan your model before you sit down at the computer. Some skill at 3D visualization and freehand sketching will come in handy here.

The point of all this discussion is that modeling is not a trivial task. In this tutorial you must try to connect the "what, how, where" of issuing commands with the "when" and "why" in order to promote your modeling goals.

We will go at quite a slow pace to start with[2]. This should leave you sufficient time for experimenting on your own, which you are strongly encouraged to do. Here's what will be covered in this lesson:

1. Starting Pro/ENGINEER
2. Layout of the screen and user interface
3. How commands are entered into Pro/ENGINEER
4. How this tutorial will represent the command sequence
5. Files and directories
6. View and display controls
7. Exploring the data structure for a part
8. Exploring the data structure for an assembly
9. How to get printed hard copy
10. How to get on-line Help

We will spend most of our time on sections 6 through 8. It will be a good idea to browse ahead through each section to get a feel for the direction we are going, before you do the lesson in detail. Even better, use the enclosed CD-ROM[3] for a multi-media overview of the lesson. A few words of caution: Take your time through the lesson. Resist the temptation to skip over the discussion and just execute the commands. There is a lot of material here which will be useful later, and not much that you can ignore without eventually paying for it. It is likely that you won't be able to absorb everything with a single quick pass-through. Good luck and have fun!

Helpful Hint

You may find it helpful to work with a partner on some of these lessons because you can help each other with the "tricky bits." Split the duties so that one person is reading the tutorial out loud while the other is doing the keyboard and mouse stuff, and then switching duties periodically. It will be handy to have two people scanning the menus for the desired commands and watching the screen. Pro/E uses a lot of visual queues to alert you to what the program is doing or requires next. Having two people watching out for this may be helpful.

---

[2] Lao Tzu also said "A journey of a thousand miles begins with a single step."

[3] The CD-ROM, by Jack Zecher at IUPUI, is included with the hard-copy version of this tutorial. For further information, contact the publisher at <www.schroff.com>.

Starting Pro/ENGINEER

To start Pro/ENGINEER on a Windows-based machine, there may be an icon right on the desktop or you may have to look in the Start menu at the bottom left of the screen on the Windows taskbar. Pro/E launches like any other Windows application. If you are running on a Unix machine (assuming the installation is standard) type ***proewildfire*** at your system prompt and press the **Enter** key[4]. The program takes a while to load so be patient. The startup is complete when your screen looks like Figure 1. The screen shown in the figure is the bare-bones, default Pro/E screen. If your system has been customized[5], the menu bars and window contents of your interface may look slightly different from this..

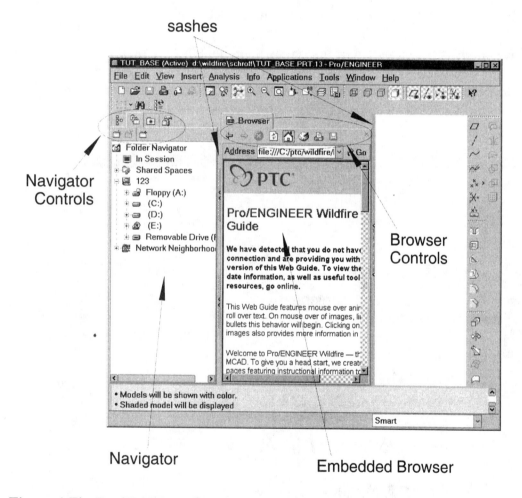

Figure 1 The Pro/ENGINEER Wildfire screen (default start-up)

---

[4] You may have to check this sequence with your local system administrator, as different installations may handle the Pro/E launch differently.

[5] Customizing the operation and interface of Pro/E is discussed in the Appendix to the hard-copy edition of this tutorial. These advanced topics are also discussed in the *Pro/ENGINEER Wildfire Advanced Tutorial* available from Schroff Development Corporation.

On the left is a multi-purpose area called the **Navigator**, with some associated controls at the top. In the default start-up, the Navigator shows you the folder structure on your machine, with the current working directory highlighted. The working directory is where Pro/E will look for and save your files. Other tools in the Navigator allow you to organize folders and web sites into groups of favorites. These are accessed using the four tabs at the top. If you place the mouse cursor over these you will see their names (**Model Tree**, **Folder Browser**, **Favorites**, **Connections**) either in a pop-up tool-tip or at the bottom of the screen. To the right of the Navigator is the **Browser**, which is an integrated web browser. A major focus of this release is connectivity between/among users. To that end, many Internet communication tools are now embedded within Pro/E. In addition to communicating with other users, the Browser allows you to launch regular web pages and, for example, download part files from the web directly into Pro/E. The browser functions in the same way as normal web browsers.

The Navigator and Browser areas can be resized by dragging left/right on the vertical sashes. Each area can be minimized by clicking on the thin textured buttons on the sashes - do that now. They will minimize along the left edge of the screen. You are left with the screen shown in Figure 2. Some shortcut buttons will not appear until we load a part.

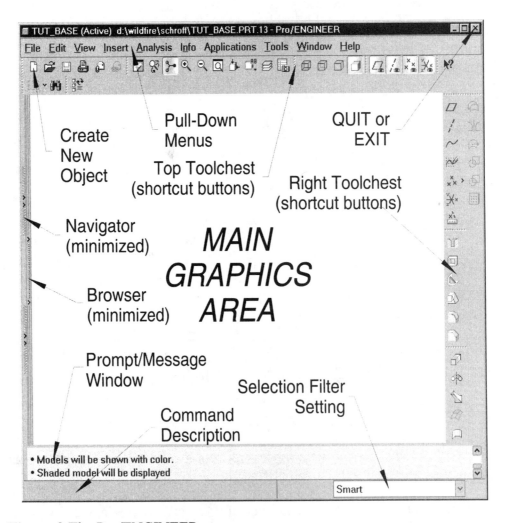

Figure 2 The Pro/ENGINEER screen

Helpful Hint

Regarding window management, DO NOT maximize the main Pro/E screen, and DO NOT resize or move the main or menu windows. Pro/E is pretty good about placing these so that they don't collide or overlap. If you start messing with the window size and placement, sooner or later you will bury a command menu behind other windows, particularly if your computer has a small screen. This will cause you a lot of confusion. Let Pro/E do its own window management for now.

In Figure 2, the main graphics area is, of course, where most of the action will take place. Users familiar with graphical interfaces (like Windows) will be quite at home with the pull-down menus at the top and the use of the short-cut buttons at the top and right side of the screen (called the *toolbars*). As you move the mouse (slowly) across the short-cut buttons or the menu items at the top, a brief (one-line) command description will appear on the bottom of the Pro/E window, and a tool tip window will pop up (this may take a couple of seconds on some systems). A number of buttons will be grayed out, meaning they are inactive at this time. The prompt/message window below the graphics area shows brief system messages (including errors and warnings) during command execution. Pro/E is usually set up to show only the last 2 lines of text in this message area, but you can resize this area by dragging on the upper horizontal border. You can also use the scroll bars at the right to review the message history. The prompt/message area is also where text or numerical input is sometimes typed in response to command prompts that ask for information (such as feature names or dimensions).

As you use Pro/E, you will encounter many other windows that will open at appropriate times. These generally act in very intuitive ways and are very similar in behavior to other Windows-based programs. In these windows, you will come across standard interface items such as radio buttons, list boxes, pull-down lists, text and/or numeric data entry fields, and buttons to accept or cancel the interaction. These windows can be moved freely around the screen, and can often be resized for convenience. When the data entered in a window is accepted, the window will generally close.

Before we load an object into the program, let's explore the interface a bit. Along the way we'll discuss how this tutorial will deal with command entry.

How commands are entered into Pro/ENGINEER

There are a number of ways that you will be interacting with the program: menu picks, shortcut buttons, keyboard entry, and special mouse functions. These are described below.

Pull-Down Menus

The main pull-down menus are presented across the top of the Pro/E window. Click on the *File* menu to open it and scan down the list of available commands. Many of these have direct analogs and similar functions to familiar Windows commands. Commands unavailable in the current context are always grayed out. The available menu choices will also change depending on the current operating mode. Move your cursor slowly across to each pull-down menu in turn (*Edit*, *View*, *Insert*, . . .) and have a quick look at the available commands. Most are grayed out at this time since we have no object loaded to work on. We will introduce these as they become available and on an "as-needed" basis as we go through the lessons. Some menu commands will open up a second level menu (these have a ▸ symbol to their right).

Short-cut Buttons

The buttons in the default toolbar immediately below the pull-down menus are shown in Figure 3. We are going to spend most of this lesson exploring the function(s) of these buttons. There are basically four button groups, as indicated on the figure. Other buttons may appear on this row as you enter different parts of the program. Buttons not relevant to the current program status are either not shown or grayed out. Move your cursor across the buttons, and a pop-up box will tell you the name of the button. The command associated with the button is described in a line of text below the graphics window. Note that there is another set of buttons on the right side of the graphics window. These are discussed a bit later. You can add your own buttons and toolbars to customize either of these areas[6].

---

[6] Customization of the interface is discussed in detail in the *Pro/ENGINEER Advanced Tutorial*. Some simple customization methods are shown in the Appendix of this tutorial. HINT: *Right click* the mouse on the toolbar.

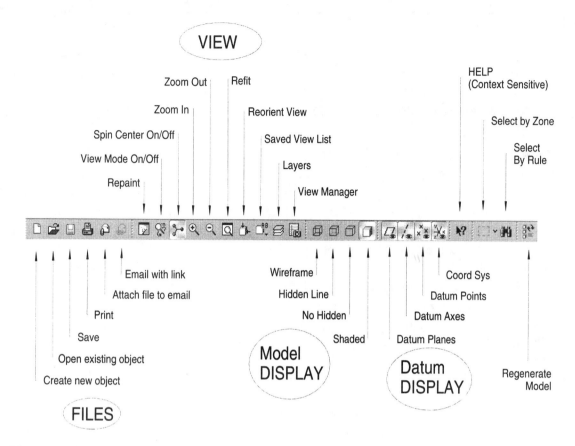

Figure 3 The top toolbar (default) with groups of related shortcut buttons

Dialog Windows

In this release, Pro/E has moved away from the cascading menu style of previous releases (a legacy of its Unix heritage), and adopted a more conventional Windows-style interface almost everywhere. Dialog windows play a major role. In these windows you must enter text or numeric data, select options from pull-down lists, set toggles (radio buttons), check options, and so on. These should be very familiar to Windows users. The idea is to set/select data in the dialog window and then select *OK*. Occasionally you will have to select *Apply* in order for the settings to "stick." You can usually set the options in any order. Sometimes, the layout and data entry fields of the dialog window will change when particular options are selected.

Menu Picks

The conversion of the Pro/E interface to eliminate cascading menus is not quite complete. There are a few places in the program where remnants of this old style still exist. In these cases, the text commands (and command options) are initiated using picks on menus that will appear at the time they are needed. These function menus will show up to the right of the main window, with commands arranged in a vertical list. As you move the mouse pointer up and down within the list, a one-line message describing the command under the pointer will appear at the bottom of the graphics window.

Helpful Hint

Each time you come to a new dialog window or menu get in the habit of quickly scanning up and down the listed commands and noting the brief message in the command window. This builds a familiarity with the location of the commands.

In the cascading menus, the pre-selected default command in a menu is highlighted. This default can often be accepted by clicking the middle mouse button (a *middle-click*) with the mouse positioned anywhere in the graphics window. With the exception of this default command, you execute a menu command by picking it using the *left* mouse button. Menu choices that are "grayed-out" are either not available on your system or are not valid commands at that particular time. Often, when you pick a command, other menus will pop open below the current one. When these represent options for the current command, the default option will again be highlighted. You can select another option by left-clicking on it. There may be several groups of options on a single menu separated by horizontal lines. Any options not currently valid are grayed out. When all the options in a menu are set the way you want (by *left-clicking*), click on **Done** at the bottom of the option menu window (or *middle-click*).

Helpful Hint

Clicking the *middle* mouse button is often synonymous with selecting **Done** or pressing the **Enter** key on the keyboard.

You can often back out of a command menu by pressing an available **Done-return** or **Quit** command, or by pressing a command on a higher menu. At some times, you will be given a chance to **Cancel** a command. This often requires an explicit confirmation, so you don't have to worry about an accidental mouse click canceling some of your work.

Pop-Up Menus

One of the big changes made recently (starting in Pro/E 2000i^2) is the number of pop-up menus used. These are available in a number of operating modes by clicking (and holding down) the *right* mouse button. This brings up a pop-up menu at the cursor location which contains currently relevant commands. These commands are often available elsewhere in the interface, but having them pop-up at the cursor location means you don't have to keep taking your attention off the graphics window. Often the alternate location for these pop-up commands is several levels deep in the menus on the side so it is much quicker to get to them using the pop-up. This is part of what Pro/E calls the "Direct Modeling" interface.

Helpful Hint
While you are learning the interface, it won't hurt to periodically execute a right-click. This will usually do no harm, and will let you get familiar with the commands available in this way. Remember that the pop-up menus are context sensitive, so they will often change depending on what you are doing!

Command Window

Occasionally, you will enter commands from the keyboard in response to prompts in the command/message window. This is also a remnant of the old interface. Generally, you will only use the keyboard to enter alphanumeric data when requested, such as object or file names, numerical values, and so on. Note that when Pro/E is expecting input in the command window, none of the menu picks will be "live."

Helpful Hint
If your mouse ever seems "dead", that is the menus, toolbars, and so on won't respond to mouse clicks, check the message window; Pro/E is probably waiting for you to type in a response.

You will have to get used to watching three areas on the screen: the menu(s) at the top and right, the graphics window, and the command/message window at the bottom. At the start, this will get a little hectic at times. Until you become very familiar with the menu picks and command sequence, keep an eye on the one-line message description in the message window. There is often enough information there to help you complete a command sequence. (Also, read the suggestion above about working with another person to start with.)

Mouse Functions

Wildfire is meant to be used with a 3-button mouse. If it has a middle scroll wheel, all the better. There are actually a number of commands that can be launched using the mouse buttons, sometimes in combination with keyboard keys. We will assume here that you have the default mouse set-up (with apologies to left handed users who may have re-mapped the mouse buttons). As you will have anticipated, most selections of menu commands, shortcut buttons, and so on, are performed by clicking with the left mouse button (LMB). In this book, whenever you "select", "click", or "pick" a command or entity, this is done with the LMB unless otherwise directed.

The functions controlling the view of the object in the graphics window are all associated with the middle mouse button (MMB) and scroll wheel (if the mouse has one). These are the important Spin, Pan, Zoom functions as shown in Table 1.1 below. Some of these are used in combination with the Shift and Control keys on the keyboard. The action of the mouse is also affected by the selection of a View Mode. All dynamic view operations involve dragging the mouse. The on-line documentation refers to this as 'Direct View Control'. The more comfortable you get with these mouse functions, the quicker you will be able to work. They will become second nature after a while. We will investigate view control functions a bit later.

Table 1-1 Common Pro/E Mouse Functions (3D)

Function		Operation	Action
Selection (click left button)		LMB	entity or command under cursor selected
Direct View Control (drag holding middle button down)		MMB	Spin
		Shift + MMB	Pan
		Ctrl + MMB (drag vertical)	Zoom
		Ctrl + MMB (drag horizontal)	Rotate around axis perpendicular to screen
		Roll MMB scroll wheel (if available)	Zoom
Pop-up Menus (click right button)		RMB with cursor over blank graphics window	launch context-sensitive pop-up menus

The dynamic view controls in Table 1-1 refer to display of 3D objects. When viewing 2D objects such as sketches and drawings, some of the mouse functions change sightly (primarily the spin command).

Mouse functions associated with the right mouse button (RMB) will also be introduced a bit later in the lessons. The main function associated with the RMB is to launch context sensitive pop-up menus as described above.

How this tutorial will represent the command sequence

This tutorial tries to present the command sequence as accurately and concisely as possible. This is a difficult task due to the different ways you will be interacting with the program, the many available shortcuts that have been added in Wildfire, and the diverse nature of the presented information (text, graphics, tool icons, line colors, menus, dialogs, etc.). Also, many functions in Wildfire have become highly automated with the use of numerous defaults. In these cases, an inadvertent mouse click on the wrong command can sometimes lead you quite far off the tutorial path. So, in the early lessons, pay very close attention and try not to jump ahead.

Pro/E generally operates on the assumption that you are an experienced user, so does not blatantly display a lot of prompting and/or unnecessary information that would slow it down. Prompts and queues that it gives are short, crisp, and sometimes quite subtle (like the color or shape of a small icon on the screen). Not much hand-holding here. And certainly no "wizards!"

We will try to discuss each new command as it is entered (usually by selecting from a toolbar or menu). Eventually, you may be told to enter a long sequence of commands that may span several menus and/or require keyboard input. Fortunately, as Wildfire adopts more and more of the standard Windows interaction methods, it is becoming easier to figure out how to tell Pro/E what you want it to do. You will know that you are beginning to understand the interface when you can enter a part of the program you have not seen before and correctly anticipate the required input.

The following notation will be used to represent command input to the program:

♦ If a command is launched using a toolbar button, that will be stated in the text, often with the button shown to help you identify it. For example: *Regenerate* .

♦ If you are to select an option from a pull-down list, you will see the name of the option with the desired list member in parentheses as follows:

Display Style(Shading)

♦ If a setting is a simple toggle, you will see the name of the option with the toggle setting (On or Off, Yes or No) in parentheses as follows:

Bell (Off)

♦ If you are to enter data through the keyboard and there is the possibility for confusion, you will see the notation using square brackets "*[...]*" as follows:

[block]

In this case, just enter the characters inside the square brackets.

♦ If you select a command (usually from the pull-down menus) that starts up another menu or window, followed by a selection from the new menu, you will see the notation using the ">" sign as follows:

menu1_command > menu2_command

♦ If a number of picks are to be made from the same menu or window you will see the notation using the "|" sign as follows (these are generally listed in a top-to-bottom order in the menu, but can be chosen in any order in dialog windows):

option1 | option2 | option3

Be aware that in some dialog windows, the contents, layout, and data entry fields may change substantially if some options are chosen.

Thus you might see a command sequence in a lesson that looks like this:

Tools > Environment > Colors(On) | Display Style(Shading) | OK

How to get On-Line Help

As you go through these lessons, you might want to consult additional reference material. Some of this might have been listed in the Browser window when you first launched the program. For the last several releases of Pro/E (since Release 18, in fact), extensive on-line help has been available. The help documentation, consisting of the entire Pro/E user manual set (many thousands of pages), are viewed using a browser. There are several ways to access the help files:

1. Selecting the *Help > Contents and Index* command.
2. Click the *What's This?* button [N?] towards the right end of the top toolbar. Then click on any command or dialog window.
3. *Right-clicking* on a command in *some* menus will show a button [Get Help] that you can press to bring up the relevant pages (context sensitive help).
4. Launch your browser and point the URL to the file[7]
                    c:/ptc/wildfire/html/usascii/proe/welcome2.htm
    or   **c:/ptc/wildfire/html/usascii/proe/default.htm**
    or   **c:/ptc/wildfire/html/usascii/proe_intro/index.htm**
    where c:/ptc/wildfire is the drive and directory where you have the program installed. Some installations may have the help files installed on a separate file or web server. Many of the help files contain links to web pages beyond your local installation (for example, the menu mapper is located on the PTC web server).

As a sampler of the extensive on-line help available, execute the following sequence:

Help > Contents and Index > Functional Area(Fundamentals)

You should see the window shown below in Figure 4. In this window, select the link *Pro/ENGINEER Fundamentals* to bring up the window shown in Figure 5. Notice the three tabs in the left pane (Figure 5): *Contents*, *Index*, *Search*. You should explore these areas a little to see what is available.

---

[7]Check this location with your local system administrator.

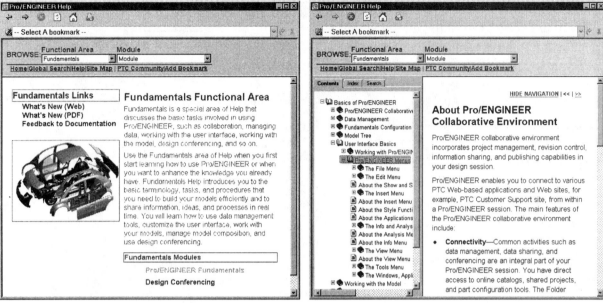

Figure 4 Opening the on-line help   **Figure 5** Pro/E Fundamentals

Under the two pull-down list fields at the top, select the link *Site Map*. This brings up a separate window shown in Figure 6.

If you have a few minutes to spare now and then, browse through the manuals (especially the *Pro/ENGINEER Fundamentals* section - see the top left of the Site Map window in Figure 6).

Another way to search by keyword is to select the link *Global Search* shown in Figures 4 and 5. This brings up the dialog window shown in Figure 7. In this window, type in a keyword at the top, select one or more functional areas listed in the left pane, move them to the right pane with the arrow, then press the **Search** button.

Figure 6 The *Fundamentals* Site Map

In the beginning, it will be a rare event when you explore the Help pages and don't pick up something useful. If you desire and have the local facilities, you can obtain hard copy of these manual pages using your browser. Your system may have postscript or pdf versions of these pages - check with your system administrator. Be aware of the cost and time involved in printing off large quantities of documentation.

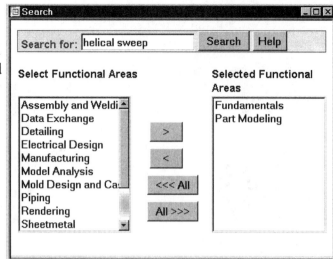

Figure 7 The Global Search dialog window

Tutorial Files and Working Directory

A number of files used in this Tutorial (parts, assemblies, drawings, and support files) are contained on the enclosed multi-media CD-ROM. You can also download an archive file (*tutparts.zip*) from the web that contains the same files; go to the site (note that this is case sensitive) <**www.schroff.com/sdcpublications/PET_download.htm**>. Finally, your instructor may have made local copies of the files available.

These files will have to be stored somewhere on your local hard disk. Note that they must all be in the same directory. This should be your default working directory for Pro/E. To find out where this working directory is, since Pro/E is up and running at this point, all we have to do is try to open a file. Pro/E will automatically look first in the current working directory. Give the following commands (starting in the pull-down menus):

> *File > Open*

or use the toolbar button shown in Figure 3 to open a file. Near the top of the dialog window (see Figure 8) the name of your current working directory is given in the *Look In* drop-down list[8]. If the dialog window looks something like Figure 8 (all the files shown in the Figure are listed in the window; you may have to scroll down the list) then proceed on to the next section on view controls. There may, of course, be more files listed in your window, depending on your local situation (other system users, computer lab common directories, and so on).

---

[8] On Windows systems, the start-up working directory can be set in the shortcut that launches the program. For a desktop shortcut, right click on the icon and select **Properties > Shortcut** and enter the path to the desired working directory in the *Start In* field.

If your *File Open* dialog window does not
contain the files listed in Figure 8, then either
they have not been installed or they are in
another directory. Depending on your
circumstance, pick from either option A or B
below.

OPTION A - Tutorial files not installed

Minimize Pro/E and copy the files to the
working directory identified above. If using the
Web download, follow the instructions on the
download page. Make sure the files are being
put in the right directory. You can then restore
Pro/E and proceed to the next section on view
controls.

**OPTION B - Tutorial files are in another
directory (changing the working directory)**

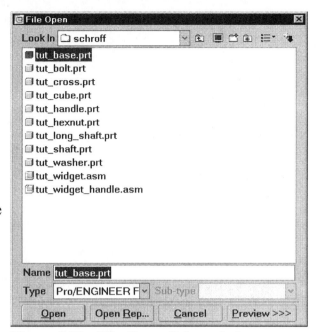

Figure 8 The **File Open** dialog window

One solution here is to copy the files from wherever they are to the current working directory.
Presumably, however, they are in the other directory for a reason. So, you have to tell Pro/E to
change its working directory by using the commands

File > Set Working Directory

then navigate using the standard Windows operations until the desired directory is shown in the
Look In field. Accept the dialog with ***OK*** (or remember that a middle click is a shortcut for
accepting a dialog window) and your working directory is now changed (see the message area).
The ***File > Open*** command should now bring up a list of the proper files.

Controlling the Screen: View and Display Commands

In this section, we will examine a number of ways to control how Pro/E displays objects on the
screen. You will need the part and assembly files described in the previous section. Although
the difference between the two main types of controls is not critical, we can distinguish between
View and *Display* controls as follows (refer back to the toolbar groupings in Figure 3):

View Controls - These let you change the orientation, size, and placement of the object in the
screen. This includes commands for zooming in and out, rotating the object, panning across an
object, or selecting a pre-defined viewing direction (like TOP, FRONT, etc).

Display Controls - These determine how the object is represented on the screen. Options here
include wireframe, hidden line, or shaded displays. All displays can be with or without color
turned on. You can also control the display style of some special line types (for example the

common edge of two surfaces that are tangent to each other - called tangent edges). There are also separate controls for the display of objects called DATUMS, which we will get to in a little while.

Let's see how these controls work.

Opening a Part File

If they aren't already, close the Browser and Navigator windows by clicking on the textured button on the right sash. Select

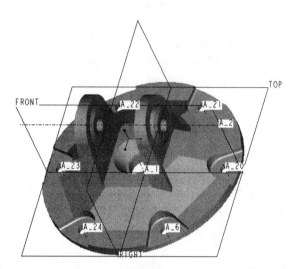

File > Open

The dialog box shown in Figure 8 will appear, showing a list of part, drawing, and assembly files in the current working directory. These can be identified by the filename extensions: **prt**, **drw**, and **asm**, respectively. Click on the file **tut_base.prt** and select *Open*. The object shown in Figure 9 should appear, possibly at a slightly different orientation. The default is a color shaded display as shown in the figure. Also shown on the

Figure 9 The part *tut_base.prt*

screen are a number of brown (or black) lines that represent non-solid geometric construction entities called *datums* (planes and axes). We will be learning lots more about datums throughout these lessons. For now, note that all datums have labels, or *tags*, like TOP, FRONT, A_1, A_6, and so on. You will also discover that datum planes have a positive side (brown) and a negative side (black). These will be clearly visible when you spin the object.

Many previously grayed out buttons on the top and right toolbars are now active. Scan these commands and buttons but resist the temptation to select any of them just yet! If you do, you can usually back out of the command with *Cancel* or *Quit* or selecting the red ✖ button at the lower right of the screen.

View Controls using the Mouse

The operation of the mouse, as far as controlling the view, depends on two related view controls: the spin center and the view mode. We'll deal with the spin center first.

Spin Center On (Default)

The display of the spin center is controlled by a button 🔗 on the top tool bar. Make sure the spin center icon in the top toolbar is on (pressed in). In the approximate center of the part you should see a small red-green-blue triad. This is the spin center. The button beside the spin center button is the View Mode control, which should be off at this time. We'll deal with it later.

The main mouse functions for view control of 3D objects are shown in Table 1-1. The dynamic view controls for spin, pan, and zoom are all performed by dragging the mouse while holding down the middle mouse button (MMB). Try the following:

SPIN   Hold down the MMB and drag the mouse. The object will spin, with the spin center staying fixed on the screen.

ZOOM   Hold down the Control key and MMB. Drag the mouse towards and away from you. This will cause your view to zoom in and out on the object. Try this with the cursor at different points on the object or screen. The center of the zoom is at the initial location of the mouse cursor.

PAN   Hold down the Shift key and MMB. Dragging the mouse now translates (pans) your view across the object.

ZOOM   If you have a mouse with a scroll wheel, try turning that to zoom in and out on the object. Once again, the zoom is centered on the initial location of the mouse when you start scrolling.

ROTATE   Hold down the Control Key and MMB. Drag the mouse left to right. The object will rotate around an axis through the spin center and perpendicular to the screen.

Spend some time experimenting with these, especially the spin command which may take some getting use to. You will be using these more than any other commands in Pro/E, so you should be very comfortable with them. See if you can obtain the three part orientations shown in Figure 10 at the right. These are the standard engineering views (top, front, right). These views are so commonly used that there is a shortcut to obtain them that we will see in a minute or two. You should also find that Pro/E will snap to these orientations when you get close to them.

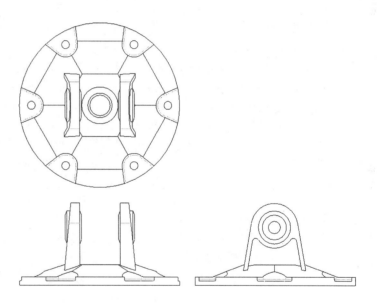

Figure 10 Standard engineering view orientations (top, front, right)

For added practice with the dynamic view controls, there is a special exercise at the end of this lesson.

Spin Center Off

Now turn the spin center off using the icon in the top toolbar and see what happens with the mouse controls. The pan and zoom controls work as before. However, using the MMB, the spin occurs around the point on the screen where the cursor is located rather than the spin center. You will find this a useful function when you are zoomed way in on the model and want to spin the part around a specific vertex. If you try this with the spin center turned on, you are liable to spin the object right out of view. So, remember that spin center OFF means that spin occurs around the cursor; spin center ON means spin occurs around the spin center. Leave it off for now as we explore the other view control button.

View Modes

The second control that works with the mouse buttons involves the View Mode. This is controlled using the icon 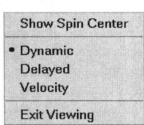 on the top toolbar. Select that now. Several changes occur on the screen. First, at the location of the spin center, a small round, black circle appears. Second, there is a black diamond with a red center. This is the current center of the dynamic view. Third, depending on the settings for your installation[9], the datum planes may disappear. If you drag the MMB, the model will spin around the red diamond on the screen. The orthogonal lines that come off the diamond form the "gnomon". The length of the main gnomon line represents how far you have moved from the initial position. While you are dragging, the diamond center turns green. When you have spun the object 90° (or 180° or 270°), the model will snap to exactly these values and the diamond will turn red. To return to the initial position, return the gnomon line to zero length. The pan and zoom functions work the same as before.

Now, hold down the right mouse button. You will get the pop-up window shown in Figure 11. Try selecting the *Velocity* option in this menu. The diamond changes to a circle. Now, drag a short distance with the MMB, and keep it pressed. You will see the object spinning at a constant speed. The speed and direction of the spin is controlled by the mouse position. Drag the mouse and the spin direction and speed will change. Spinning stops when you release the mouse button. When you are in velocity mode, **Figure 11** The View both pan and zoom will occur at a constant velocity until you release the Mode pop-up menu mouse button.

Hold down the RMB and select the Delayed option. The circle changes to a square. Any view changes are delayed until the mouse button is released. One problem with delayed mode is it is very easy to make changes that are too large (especially pan) and when the view is updated the object is completely off the screen. If that happens, use the *Refit* button in the top toolbar (Figure 3). Note that this cancels View Mode, as does pressing any of the datum display buttons.

As you can see, there are a lot of visual queues to alert you to the current view control state.

The black circle means that View Mode is on. You can turn it off either using the top toolbar icon or select *Exit Viewing* in the pop-up menu (Figure 11).

If the spin center is turned back on, the view mode operations still work as we have just seen. In this case, though, spinning occurs relative to the spin center, as you might expect!

---

[9] There is a configuration setting (see the Appendix) that determines whether the datum planes are displayed during some operations. The option is *spin_with_part_entities*. When this is off (default) the datums will disappear whenever you spin the model.

Toolbar View Commands

There are several other buttons on the top toolbar to control the view. The most important of these is probably the ***Repaint*** button, that looks like ⬚ . You can also use ***View > Repaint*** or simply press CTRL-R (hold the Control key while you press R). This command causes a complete refresh of the graphics window, which is sometimes helpful to remove entities no longer required for display (like feature dimensions). Try out the other view buttons in this group ⊕ ⊖ ▢ (***Zoom In***, ***Zoom Out***, ***Refit***), watching the message area for prompts. The handiest of these is probably the one on the right (***Refit***).

Using Named Views

In addition to the dynamic viewing capabilities available with the mouse, you can go to predefined orientations. To view the object in the default orientation, select the ***Saved View List*** shortcut button ⬚ in the top toolbar and click on ***Default***. Alternatively, you can select

View > Orientation > Default

or press ***CTRL-D*** (hold the Control key while you press D). Your screen should now look like Figure 9 above. Try the other named orientations (TOP, FRONT, and so on) in the saved view list. When a drawing is made of a part or assembly, these are exactly the views that most commonly appear. You should bear this in mind when you are creating parts - orientation matters! In the named views (TOP, FRONT, RIGHT) you are looking directly at the brown side of the datum plane of the same name. It appears as a rectangle around the part. This rectangle will stretch or shrink so that it always just encloses the part. The other datums in these views are seen edge-on as pairs of brown and black lines. As mentioned above, the negative side of a datum plane is indicated using black. It is important to get some practice with looking at datum planes and, from their color, figuring out your view orientation.

Most parts you create will have these standard engineering views already defined, using standard part templates which we will discuss later. If you want to experiment with creating your own named views (handy for documenting the model), select the commands

View > Orientation > Reorient

(or use the ***Reorient View*** shortcut button ⬚) which brings up the dialog shown in Figure 12. To expand the lower half of the window, click on the blue "Saved Views" region of the window. These named views have already been created in the part. In fact, they will be automatically present in any parts you create using the default template, discussed in the next lesson. If you want to set up your own named views, the general procedure for the ***Orient by Reference*** type (selected in the pull-down list at the top of the window) is to select two orthogonal surfaces or datum planes and tell Pro/E which way they should face in the desired view. These are called the view references and are specified in the **Options** area of the dialog window. References can be chosen to face the top, front, left, right, front, or back of the screen. For the example shown in Figure 12, Reference 1 is chosen as **Front**, that is the reference will face the front of the screen (toward you), while Reference 2 has been set to **Top** (the reference will face the top of the screen). In Figure 12, Reference 1 is the RIGHT datum plane; Reference 2 is the TOP datum plane. This combination of planes and directions produces the standard RIGHT engineering view (the saved name in the pane at the bottom). Note that you could obtain the same orientation by picking Reference 2 as the FRONT datum plane facing the left edge of the screen.

Once the view orientation options have been selected, you can enter a new view name at the bottom, then select **Save**. If you select a view in the **Saved Views** list, then **Set**, the model will spin to that orientation, but the options shown at the top of this dialog window are not updated. Data in this dialog window flows top-down only.

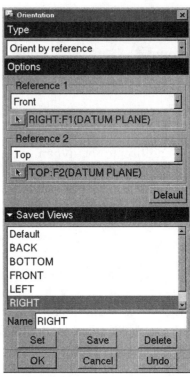

No doubt this all seems terribly confusing since we are using the same words ("top", "front", "right") for three purposes: naming the datum planes, specifying view directions relative to the screen, and giving names to the saved views themselves. Rest assured that this will eventually start to make sense! Close the Orientation window with **OK** (if you have made changes you want to keep) or **Cancel**.

Figure 13 Dialog for creating saved views

Object Display Commands

These buttons on the top toolbar are fairly self-explanatory. They are (see Figure 3):

Wireframe - all object edges shown as visible, even those at the back
Hidden line - hidden edges (or portions) displayed in gray (dark but visible)
No Hidden - hidden edges (or portions) not shown
Shading - shaded solid model

These display modes persist until changed by selecting a different mode. Try them out now. These buttons do not affect the display of datums. The datums may disappear momentarily when you are changing the view, for example during spinning. Note that the command

> *View > Shade*

produces the same immediate effect as the **Shading** button except that the datums are automatically turned off and the display mode persists only until the next **Repaint** regardless of the display toolbar buttons. Of these four modes, you will probably spend most of your time in hidden line mode, since this allows you to see hidden features (although they are dimmed). If you have a slow graphics card, this mode is also quicker than shaded images. Slow graphics cards are especially noticeable during spinning of shaded objects, which can become quite jerky for complicated models (single parts with lots of surfaces, or assemblies of many parts). However, shaded mode is useful to reinforce your mental image of the three dimensional shape, when that is sometimes unclear by looking at only the edges. The display mode is very much a matter of personal preference.

Datum Display Commands

The four buttons in the DATUMS group of the top toolbar control the visibility (on or off) of the datums (planes, axes, points, coordinate systems). Try these out now. This model does not contain any datum points or coordinate systems, so you won't notice any effect from those two buttons. Remember that these buttons do not delete these entities from the model, just turn off their display. Also, do not confuse the datum display buttons at the top (each with a little eyeball on them) with the datum creation buttons (without the eyeball) on the right toolbar.

You can also control the display of the datum labels, or tags. Select

View > Display Settings > Datum Display

See Figure 13. Turn off all the tag display settings. Accept the dialog with OK. The labels are gone[10]. Turn them back on again with the same sequence.

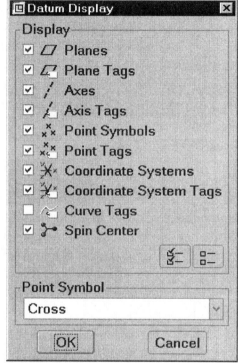

Figure 13 Datum display settings

_____

[10] You can easily add shortcut buttons to the top toolbar to control the display of datum tags. See the Appendix section on screen customization.

Modifying the Environment

Try experimenting with the settings in the dialog window obtained using

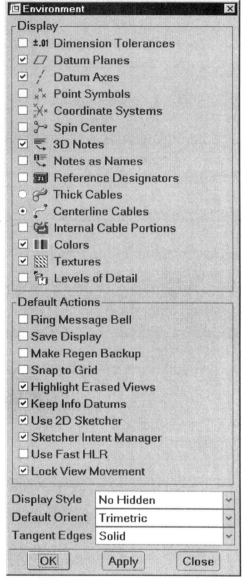

Tools > Environment

See Figure 14 at the right. This single window allows you to set the display of datums, the spin center, colors, as well as the overall display style, orientation, and treatment of tangent edges. Near the bottom of this window, try setting the display style to *No Hidden* and tangent edges to *Dimmed*. Now select *Apply*. Observe the screen carefully to see where the tangent edges have appeared. (You should be aware of tangent edges when you are modeling, since it is not good practice to use them as feature creation references.)

Your new settings will take effect when you select *Apply* or leave the *Environment* menu with *OK*. What happens if you make changes in the settings and then select *Close*? Note that the most common display styles are easily obtained using the short-cut buttons in the top toolbar.

Another common setting involves the Colors option. Set the following:

Colors(Off)
Display Style(Shading)

then select *Apply*. This shows the part in the default neutral gray color. If you set the display style to *Hidden*, you will see the visible edges as white lines, hidden edges as gray lines, and tangent edges as dark gray lines.

Figure 14 The Environment settings dialog window

Your choice of display style is strictly personal preference. However, keep in mind the following:

▸   It is useful to be able to see the entire object at once. *Hidden Line* gives you "x-ray" vision to do just that. *No Hidden* hides too much, and *Wireframe* is sometimes ambiguous.
▸   A lot of information is portrayed using color queues. It is useful, then, to keep your model color neutral so that you won't miss anything.
▸   On the other hand, the 3D shape of the object is most clearly shown in a shaded image. If you have trouble interpreting the line drawing image, go ahead and shade the part. And, color adds interest to what you're doing!

For now, turn *Colors* off, and set the display style to *Shading*.

Anatomy of a Part - Understanding the Model Structure

Now that we have explored most of the viewing and display commands, let's explore the model itself[11]. If the part **tut_base.prt** is not loaded, do that now with ***File > Open***. Open the Navigator pane and select the left-most control tab (***Model Tree***). The screen should look like Figure 15 below. Let's examine the information in the model tree in the left pane. This contains a list of the features in the part database. The features were created in a top-down order in this list, called the *regeneration sequence*. Each of these features has a name - some are the default names and others have been specified by the model creator. Beside each feature is a small icon to help you identify the type of feature. We will discover more about these icons in a couple of minutes. As you click on any of the features listed in the model tree, edges of the feature are highlighted in red on the part in the graphics window. Try that now.

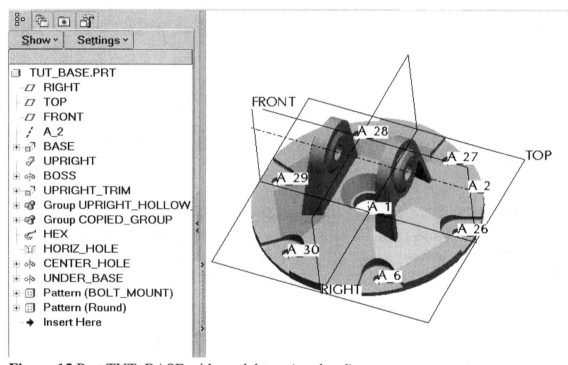

Figure 15 Part TUT_BASE with model tree (top level)

Preselection Highlighting

Preselection highlighting is how Pro/E tells you what will be selected on the model if you pick with the left mouse button. This tool was introduced in Release 2001 as part of the new object/action command structure being introduced in that release. Pro/E previously allowed only the action/object style of command (and still does). In the newer object/action structure, you select an object first and then specify the action to be performed on it. For many people, this is a more natural way of doing things. To accommodate the new mode, Pro/E allows you to pick

---

[11] It's taken a while, but this is where you finally get to see the fish!

most objects directly in the graphics window by clicking on them[12]. When the display gets very cluttered, it may be hard to pick on exactly what you want with the first click of the mouse. Thus, preselection highlighting is a method to make sure you are picking what you want in a complicated model.

To see how preselection highlighting works[13], first make sure that **Smart** is selected in the **Selection** filter list at the bottom right of the screen, as in Figure 16.

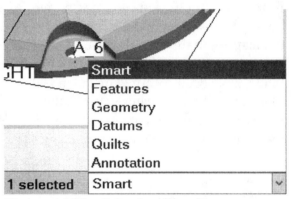

Place your cursor on one of the surfaces of the model, but do not click just yet. Some edges of the model will highlight in cyan and a small box will pop open beside the cursor (there might be a couple of seconds delay). The box and highlighted edges show you which feature is involved with or **Figure 16** The Selection filter list has created the highlighted surface. As you drag the cursor (slowly!) around on the model, the different features are identified that have created different surfaces.

If a feature is highlighted and you pick on it with the left button, the highlighted edges will turn red and the feature name will be highlighted in the model tree. The identified feature is now selected as the object to be acted upon by the next command. To go deeper into the geometry of that feature, slowly move the cursor over the highlighted (red) surfaces. Now the filter will identify individual surfaces and edges formed by the feature. If you left click on an edge, you can go down to the next level and pick an individual vertex. The entity currently highlighted in red at any time during this selection process is the active entity. As soon as you have the one you want, you can leave selection and proceed on to perform some action on this entity (for example by picking a command from the RMB pop-up menu).

If you set the selection filter to (**Datums**), only datum planes and axes are trapped by the filter. The selection filter (**Geometry**) is the most detailed - it traps surfaces, edges and vertices and shows who they belong to.

Expanding the Model Tree

We would like to set up the model tree to show a bit more information. We will find out how to do this from scratch in a later lesson. For now, we are going to read in a model tree configuration file that has been created for you. This should be in your working directory, along with all the

---

[12] A function called **Query Select** has always been available to do this as part of the action/object selection process, and we will investigate it a bit later. The new preselection method is much simpler.

[13] Make sure this is turned on using **Edit > Select > Preferences**, and make sure the option **Preselection Highlighting** is turned on (checked). This is the Pro/E default.

part files transferred from the CD-ROM or downloaded from the web. To load it, select the **Settings** tab at the top of the model tree. Then pick

Open Settings File

Select the file **mod_tree.cfg** and click ***Open***. This will add some columns on the right of the model tree (see Figure 17). You may have to drag the right border out a bit to see all the columns. The columns can all be resized by dragging on the vertical separators at the top. The new columns are the feature number, feature type, and feature subtype. The feature number indicates the order of feature creation (*regeneration sequence*). The feature types are datum planes, protrusions, cuts, holes, and so on. The subtypes give further classification information about each feature. For example, feature #5 is an extruded protrusion, feature #6 is a parallel blend protrusion, feature #7 is a revolved protrusion, etc. We will be spending

		Feat #	Feat Type	Feat Subtype
▢	TUT_BASE.P			
	RIGHT	1	Datum Plane	
	TOP	2	Datum Plane	
	FRONT	3	Datum Plane	
	A_2	4	Datum Axis	
+	BASE	5	Protrusion	Extrude
	UPRIGHT	6	Protrusion	Blend, Parallel, R...
+	BOSS	7	Protrusion	Revolve
+	UPRIGHT_	8	Cut	Extrude
+	Group UPF	9	Group Head	
+	Group COF	12	Group Head	
	HEX	18	Protrusion	Sweep
	HORIZ_HC	19	Hole	
+	CENTER_I	20	Cut	Revolve
+	UNDER_B	21	Cut	Revolve
+	Pattern (BC	22	Pattern	PATTERN
+	Pattern (Rc	53	Pattern	PATTERN
➔	Insert Here			

Figure 17 Model tree with added columns

a lot of time in these tutorials finding out what these features are and how they are created.

Most of the features in this part have been named. This is optional, but a very good idea. Features that have not been named will appear with only the feature type (protrusion, cut, hole, etc.) in the first column of the model tree. In a large model, this becomes very confusing and not very helpful. You should get in the habit of naming at least the key features in your models.

Notice that there seems to be a few features missing (notice the gaps at 9 - 12, 12 - 18, and 22 - 53). Click on the small plus sign beside feature #12 (on the left). This will expand the model tree to the next level for this feature, revealing several features there. Although these have not been named, they are members of a named group. The group was formed by mirroring features 6 through 11 simultaneously through the RIGHT datum plane. The group name indicates the mirror operation.

Open the group starting at feature #9. In this group, the icon beside feature #10 (DTM5) is in a gray box. Notice that this datum plane is not visible on the model - it is currently hidden (as indicated by the gray box icon). If you put the cursor on DTM5 in the model tree and hold down the right mouse button, you can select ***Unhide***. The datum plane now appears. This plane was used to create the cross sectional sketched shape of the next feature (#10, extruded cut). If you press the + sign beside feature #10, you will see a feature S2D0002. This is the sketch that defines the shape of the cut. Highlight this and hold down the right mouse button. Select ***Edit***. The actual sketch now appears in yellow. It is best seen in RIGHT view. Select DTM5 again and *Hide* it (using the RMB pop-up menu).

Close the group at feature #9. If there is a bunch of extra stuff on the screen, use ***Repaint***. Find out what is in the Pattern starting at feature #22 by expanding the model tree. Technically, this is called a *radial pattern* since all instances of the pattern were created by incrementing an angle around a central axis. The pattern instances are groups of features - these are named in the first group. Each group has two hidden datums. See if you can find out what they are for.

The last pattern consists entirely of rounds. These rounded edges were kept separate from the previous group pattern because rounds are normally considered cosmetic features. Rounds often cause problems with downstream operations such as creating drawings and performing finite element analysis. Adding them last, and keeping the rounds separate means they can be easily selected and temporarily·removed from the model (or *suppressed*, see Lesson #5).

Compare the feature types and subtypes listed in the model tree with the feature icons on the left. You should be able to easily identify datum planes and axes, and extruded, revolved, and swept features, patterns and groups, and hidden features from their icons.

The Model Player

Just knowing the feature types and creation order doesn't give you a really clear idea of how this model was created. A function introduced in Release 2001, called the ***Model Player***, offers a useful tool to investigate this. This is handy if (like now!) you have obtained a model and want to explore its structure.

Make the model tree a bit smaller so that all you can see are the feature names and numbers. Now select

Tools > Model Player

A new dialog window opens (Figure 18). This operates much like a tape player, with buttons at the top to let you move forward and backward through the regeneration sequence. Let's start by picking on the button on the far left to get to the beginning of the sequence. The display shows we are feature 0, with nothing shown on the screen. The model tree shows <none> beside each feature.

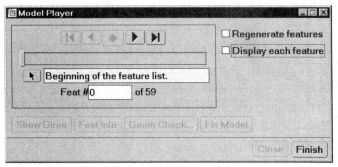

Figure 18 The Model Player window

Exploring the Structure of a Part

We're going to step through the creation of this part using the Model Player. This should give you a feel for how models are created and how the features work together to create the geometry. This very simple part illustrates quite a number of interesting and important ideas about modeling with Pro/E.

In the Model Player, click on the button (second from right) to step forward a single feature. This brings feature #1 into the model (the RIGHT datum plane). Continue clicking this button until you are at feature #5. This is the first solid feature of the part. It is an *extruded protrusion*, created by sketching a circle on the TOP datum plane and extruding it. A *protrusion* adds solid material to a model. (A *cut* takes it away.)

Click on the **Show Dims** button in the Model Player window. A yellow circle and a couple of dimensions appear on the model (Figure 19). This was the sketch used to define the shape. The circle has a diameter (note the symbol Ø) of 200. The sketch was then extruded a distance 10 to form the solid feature. These are the *parametric dimensions* of the feature.

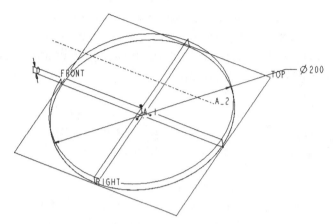

Figure 19 Feature #5 with **Show Dims**

Move on to feature #6, the UPRIGHT and turn off the datum planes. Go to the Top view and select **Show Dims** in the Model Player. You will see all the dimensions of the sketch used to create this *blend* (containing two rectangles). The blend feature allows you to specify different cross sections along the length of a protrusion. This lets the upright taper towards the top - blending from the large rectangle at the bottom to the small one at the top. We will deal with blends later in these lessons - they are quite an advanced feature and can do much more than demonstrated here.

Still on feature #6, click the **Feat Info** button. This opens the Browser pane and presents a great deal of information about the feature: its feature number, an internal ID number, the number and ID of parent and child features, all its attributes and dimensions, and so on. This information will make more sense to you as we proceed through the lessons. The concept of parents and children is quite important. A *parent* feature is one that supplies references necessary for the creation of a subsequent feature, which then becomes a *child* feature of the parent. The circular disk (feat #5, BASE) is a parent for the UPRIGHT (#6), because its upper surface was the sketching plane for the blend. The BOSS (feat #7) is a child of the UPRIGHT since one of the UPRIGHT's surfaces was used as a reference for the BOSS's sketched shape. **Close** the Browser pane by picking the button on the right sash.

In the Model Player, continue on to feature #7, the BOSS. This is a revolved protrusion, created by sketching a 2D shape on the FRONT datum plane and revolving it around axis A_2. This was the reason for creating this datum axis - an example of planning ahead. Go to the default view, turn on the datum planes, click on **Show Dims** and zoom in on the model to see the sketch.

Move on to feature #8 - this is an extruded cut that trims the top off the upright. *Cut* features always remove material (the opposite of a protrusion). Click on **Show Dims**. The sketch for this feature is on the RIGHT datum plane and has no dimensions! All its geometric information is picked up from previously created features (the sides and top of the upright).

Proceed to feature #11, which is another extruded cut used to hollow out the upright. Click on

Show Dims and zoom in to see the sketch for this feature. What plane is it sketched on? This is tricky - there was no surface available to define the sketch at the right distance (5 units) from the left vertical surface of the UPRIGHT. So, a special datum plane was used. In previous releases, this would be called a *make datum* (or a *datum-on-the-fly*)[14]. This appears as DTM5 in the model tree. Once the sketch was defined, the make datum disappears from the display (but is listed in the model tree as a hidden feature). Make datums are very powerful tools and we'll find out more about them in lesson 5.

Features #6 through #11 are now mirrored through the RIGHT datum plane to create the other upright. The mirrored features form a group, since they were all mirrored at the same time. Note the group name in the model tree.

Feature #18, HEX, is an interesting one. It could be made with a very shallow blend between two hexagons, in the same way as the UPRIGHT was made. HEX was, however, made in a different way using a *sweep* (observe the icon on the model tree). A swept feature requires a cross section shape that is then moved along a trajectory to sweep out a solid. HEX was created by sweeping a single inclined edge along a hexagonal trajectory. You can see these if you select ***Show Dims***. The sweep trajectory is a regular hexagon inscribed in a construction circle of radius 90 (note the "R"). The swept "section" consists of a single line with one end touching the hexagon. A number of special options have been set in creating this feature so that this single line creates solid geometry instead of a surface as it is swept. Sweeps can be used to create cuts or protrusions. They come in many varieties and are among the most complex features to create in Pro/E. We will look at the simpler versions later in this tutorial[15].

Why, do you suppose, was the feature HEX made *after* the two uprights in this model? This feature creation order is important. (HINT: consider the cut feature UPRIGHT_HOLLOW)

The next feature (#19) is the horizontal hole going through the two bosses. This hole is defined (use ***Feat Info*** again) on the RIGHT datum plane, is coaxial with A_2, and extends through everything in both directions (technically called a *Thru All, Both Sides* hole). The only dimension required is its diameter.

Features #20 and #21 are both revolved cuts in the bottom of the part. It would be technically possible to combine both these features into a single revolved cut. This would make the sketched shape more complicated (and hence more difficult to create) and would also give less flexibility to the model. For example, feature #21 (UNDER_BASE) can be easily deleted from the model without affecting feature #20 (the CENTER_HOLE). If these were both parts of the same revolved feature, this would be more difficult. When choosing the features to make up a model, there is always the question of maintaining a balance between a large number of simple features and a smaller number of more complicated ones. When first starting out in Pro/E it is probably

---

[14] It appears from some documentation that PTC now wants to call these *asynchronous datums*. Kind of a mouthful!

[15] An entire lesson in the *Advanced Tutorial* is devoted to more advanced versions: helical sweeps and variable section sweeps.

better to lean toward the former approach. This is what we'll do in this tutorial.

Looking in the model tree, observe that immediately below feature #21 there is an entry for a *pattern*. Expanding this to the next level shows a number of groups. Expanding the first group (#23) shows four features: a hole, a couple of hidden datums, and a cut. Left click on each of these to see them highlighted in the model. The group is then duplicated around the circumference of the base. This is called a *radial pattern* which is produced by incrementing an angular dimension parameter associated with the first group (the pattern leader). There are many other different ways of forming patterns - we will see several of these later in lesson 7.

Advance the model player all the way to the end of this pattern by enter **52** in the **Feat #** text box in the model player. This is the last feature in the last group of the pattern.

This almost concludes the regeneration sequence of the part using the model player. Note that you can step backwards through the sequence, or enter the desired feature number directly in the window. There are some other optional functions that you can explore on your own.

The last six features in the part are rounds. Bring them into the model, and leave the model player, by pressing the ***Finish*** button.

Close up all the levels in the model tree by clicking on the small minus signs, so that only the first level is showing as in Figure 17.

Modifying Dimensions

There is lots more we can do with the information in the model tree. Here are three ways you can modify the dimensions of a feature.

First, select feature #5 (BASE) in the model tree. Hold down the right mouse button with the cursor on the feature name. A pop-up menu appears. Select ***Edit***. The feature sketch and dimensions appear in yellow on the screen. Double-click directly on the thickness dimension (currently 10). A small window will appear at the dimension location. Type in **25** and press Enter. Nothing happens on the screen except that the dimension now shows in green (the diameter dimension is still yellow). A green dimension indicates that although the value of the parameter is changed, the geometry has not been updated to reflect the new value. To do this, we must *regenerate* the model. The command to do this is either (starting in the pull-down menu):

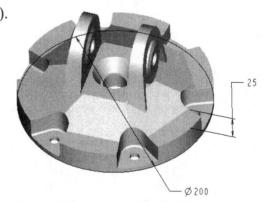

Figure 20 Part with modified base thickness

Edit > Regenerate

or (this is quicker) use the shortcut button ⧉ at the right end of the toolbar at the top. Observe the message window. Eventually, you should be informed that the regeneration was successful. Look at the thickness of the model. See Figure 20.

Let's change the thickness again, but using a different way to launch the command. Make sure the selection filter is set to **Smart**. Move the cursor across the model until the BASE feature is preselected (cyan). Then left click to select it (it highlights in red). Now right click to bring up a pop-up menu in the graphics window. In this menu, select *Edit*.

What happens if we make a mistake here? Change the thickness of the BASE from 25 to *50*. Can you predict what will happen? Once again, you will have to select the *Regenerate* command in the PART menu. This time, the regeneration will fail on feature #11 (UPRIGHT_HOLLOW). There are some tools available in the RESOLVE FEAT (Resolve Feature) menu at the right to try to discover exactly why this feature failed[16]. We will be looking at those tools in lesson 5. For now, select *Undo Changes > Confirm*. This returns the thickness back to 25.

A final way to change the value of a dimension is very easy. Just double-click on any surface of the base feature. Make sure you don't accidentally pick a surface of HEX or one of the cuts. If you get the disk, the sketch will appear with dimensions. Change the thickness dimension back to *10* and regenerate the model. Double-clicking on any sketched feature will bring up its sketch and dimension values - no need to go looking for a menu command.

Parent/Child Relations

Select HEX (either in the model tree or using preselection in the graphics window), right click and select *Info > Parent/Child*. This brings up the reference information window shown in Figure 21. The lower left pane shows all the parents of HEX and the right pane shows all its children. Clicking on any of these features will highlight them on the model.

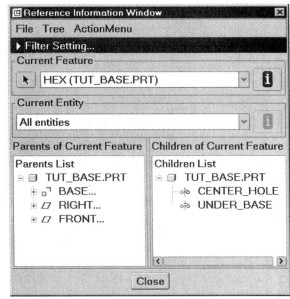

The feature BASE is a parent of HEX. Click the + sign beside BASE in the Parents pane. This identifies the surface of BASE that was used as a reference for HEX. How? Click on the surface entry in the pane and read the message window. This tells you that the surface (shown in pink) was the sketching plane for HEX.

Figure 21 Exploring Parent/Child relations

What other children does BASE have? If you select BASE in the parents list and hold down the right mouse button, another pop-up menu appears. In this menu, select *Set Current*. This changes focus to BASE (look at the Current Feature field at the top of the window). The child

---

[16] The change in base thickness has caused the sketch for the UPRIGHT_HOLLOW feature to cross through itself - this is not allowed!

list contains HEX and numerous other features.

The relations between parents and children can become very complicated. It is crucial when starting a new model to carefully consider how these are going to be set up - which features should depend (and how) on which others, and which should be independent. Insufficient planning here (or more usually none at all!) will result in a lot of grief later when the model needs to be modified.

This concludes our exploration of the model of a single part. Close the reference information window and the model tree.

Now, of course, very few parts exist in isolation from one another. Most parts are used in assemblies that must perform some function. Let's move on to an assembly and spend some time exploring how it is set up.

Anatomy of an Assembly

We will load an assembly that uses the base part we have just been examining. Set up the screen as follows: shaded color display with datums off (they just clutter up the view!). However, be aware that since color plays a significant role in determining the meaning of lines, it is usually best to avoid color displays unless you have a really good reason not to (like preparing images for documentation). You will need to have all the tutorial files installed as discussed earlier. Select the second tab (*Folder Browser*) in the Navigator window. This opens the Folder Browser that lists the directory/file contents of your local disk. Select the directory where you have installed the tutorial files. Open the Browser window. This now shows a listing of files in the chosen directory. Even better, if you pick on one of the part or assembly files, a preview image appears at the top of the Browser window (Figure 22). The usual dynamic view controls work with this window.

Before we proceed, observe the size of the file **tut_widget.asm**. This is the assembly file we will open in a minute. Compare its size to the part file **tut_base.prt** that we have been examining. The assembly file is quite considerably smaller. Why? The reason is that the assembly file contains only the necessary information for *how* the parts are put together in the assembly. *The assembly file does not contain the parts themselves.* The most common cause of an assembly error is not having the appropriate part files available (where Pro/E can find them). This is a

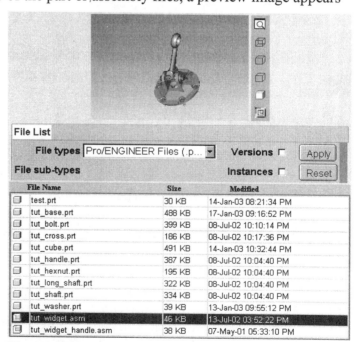

File Name	Size	Modified
test.prt	30 KB	14-Jan-03 08:21:34 PM
tut_base.prt	488 KB	17-Jan-03 09:16:52 PM
tut_bolt.prt	399 KB	08-Jul-02 10:10:14 PM
tut_cross.prt	186 KB	08-Jul-02 10:17:36 PM
tut_cube.prt	491 KB	14-Jan-03 10:32:44 PM
tut_handle.prt	387 KB	08-Jul-02 10:04:40 PM
tut_hexnut.prt	195 KB	08-Jul-02 10:04:40 PM
tut_long_shaft.prt	322 KB	08-Jul-02 10:04:40 PM
tut_shaft.prt	334 KB	08-Jul-02 10:04:40 PM
tut_washer.prt	39 KB	13-Jan-03 09:55:12 PM
tut_widget.asm	46 KB	13-Jul-02 03:52:22 PM
tut_widget_handle.asm	38 KB	07-May-01 05:33:10 PM

Figure 22 Browser with preview window (Navigator tab *Folder Browser*)

common mistake that newcomers make when transferring work from one computer to another, thinking that all they need is the assembly file itself.

Select the file **tut_widget.asm** in the Browser and select the lowest button in the set to the right of the preview image. This will *Open* the file. It may take a few seconds to bring in the assembly. See Figures 23 and 24.

If the selection filter is set to *Smart*, moving the cursor back and forth across the assembly will bring up pop-up boxes that list the part names. The parts in Pro/E are called *components*.

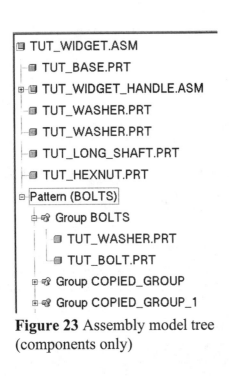

Figure 23 Assembly model tree (components only)

Figure 24 The assembly **TUT_WIDGET**

You can see that the model tree is laid out in the same way as features in a part. Clicking on a component in the model tree will cause it to highlight in dark red on the model. Note the icons in the model tree. The second component is itself an assembly - usually called a sub-assembly. Opening the model tree of the sub-assembly to the next level shows that it is made up of six other components. Sub-assemblies can be nested like this to as many levels as you want, and are a good way to organize the structure of a complicated assembly. Different groups in a company can even work on different sub-assemblies at the same time, although file management does become an issue for very complex models.

Some parts (like TUT_WASHER) are used several times in the assembly (10 in fact), but only a single washer part file is on the hard disk. This shows that components can be re-used multiple times within an assembly. Common components, like fasteners, can be used in many different assemblies. This can obviously save a lot of file storage space. In fact, many companies will maintain a library of commonly used parts.

Notice at the bottom of the model tree is a pattern. The pattern leader is a group (BOLTS) of two components (TUT_WASHER and TUT_BOLT). Each of the other groups in the pattern contains these components. This is similar to the structure of a pattern of grouped features.

An Assembly BOM (Bill of Materials)

Here is a very useful (and necessary) tool for assemblies. In the pull-down menus, select

Info > Bill of Materials

In the BOM window, select **OK**. This opens the Browser window with a BOM report for the assembly. This report contains three tables, the last one is shown in Figure 25. The buttons in the three right columns in each table allow you to highlight the part in the model display, get model info, or open the individual component. If desired, you can use the Browser controls to save or print this BOM. In any case, the BOM data is automatically saved to a file (in this case, *tut_widget.bom*) in the current working directory.

Quantity	Type	Name	Actions
1	Part	TUT_BASE	
1	Part	TUT_CROSS	
1	Part	TUT_HANDLE	
10	Part	TUT_WASHER	
1	Part	TUT_SHAFT	
2	Part	TUT_HEXNUT	
1	Part	TUT_LONG_SHAFT	
6	Part	TUT_BOLT	

Summary of parts for assembly TUT_WIDGET:

Figure 25 Bill of Materials summary table

Modifying the Assembly

Returning to the model tree in the Navigator, let's add the extra columns in the model tree as we did for the single part. Select the tab **Settings,** then **Open Settings File** and select the file **mod_tree.cfg**. The extra columns are added (Figure 26).

The model tree now also shows what are called *assembly features* in the model. For example, the first three entries on the model tree of the assembly are datum planes. Datum features are very common in assemblies, since they provide references for placing and constraining components. To see an example of this, select feature #5 **ADTM1** in the assembly model tree (the datum plane does not have to be visible on the screen), right click, and select **Edit**. An angular dimension will appear in yellow

	Feat #	Feat Type	Feat Subtype
TUT_WIDGET.ASM			
ASM_RIGHT	1	Datum Plane	
ASM_TOP	2	Datum Plane	
ASM_FRONT	3	Datum Plane	
TUT_BASE.PRT	4	Component	
ADTM1	5	Datum Plane	
TUT_WIDGET_HAND	6	Component	
TUT_WASHER.PRT	7	Component	
TUT_WASHER.PRT	8	Component	
TUT_LONG_SHAFT.F	9	Component	
TUT_HEXNUT.PRT	10	Component	
Pattern (BOLT_WASH	11	Pattern	PATTERN
Insert Here			

Figure 26 Assembly model tree using *mod_tree.cfg*

that gives the rotation of this datum plane around the horizontal axis **A_2** in the base part. Set the RIGHT view to see this clearly. Change this angle by double-clicking on the number and typing in a new value, say **90**, then press Enter. The value will stay green until the model is regenerated.

Notice there are now two **Regenerate** toolbar icons on the top toolbar. The one on the left is the same as before, and will regenerate the entire model. This can be time consuming for very large models. If you are sure the regeneration should affect only a limited number of components, try the **Custom Regenerate** button on the right. This won't act much differently in this assembly because the feature we are changing (ADTM1) occurs so early in the assembly sequence. Everything after it is affected if it is modified.

The assembly will regenerate with the subassembly positioned at a new angle (Figure 27). The datum plane **ADTM1** in the top assembly serves as an orientation reference for the subassembly. When the datum plane moves, so does the subassembly. Repeat the **Edit** command to put the subassembly back in its original position (change the 90 back to **30**).

Figure 27 Modified assembly

In the model tree, expand the sub-assembly TUT_WIDGET_HANDLE. It also contains an assembly datum ADTM1 which controls another angle. Try changing this to 30° and regenerating the model. Clearly, there is a problem with the model - at least two parts (the handle and the base) are interfering with each other. This interference is quite obvious. Later in these lessons we will find a Pro/E command that will identify interference automatically, even ones that you don't expect. For now, put the datum ADTM1 in the sub-assembly back in the original position (75°) before you continue.

In addition to assembly datums, you can even create regular features like cuts and protrusions in the assembly that act on the components. An example of a cut feature would be to produce a cutaway view of an assembly.

Notice that each component listed in the model tree can be expanded to show its individual features: click on the + sign to the left of the component icon on the model tree. For example, Figure 28 shows the expansion of component #10 TUT_HEXNUT. Selecting part features allows them to be modified the same as before. Try changing the thickness of the base part as we did before. What happens to the bolts and washers?

You can get the same sorts of information about components in the assembly (component info and parent/child relations) as you could for a part. These are usually launched, as before, with a right mouse click in the model tree. In the case of an assembly, the parent/child relations of components involve the datums, surfaces, and edges of components that are used

	Feat #	Feat Type	Feat Subtype
TUT_WIDGET.ASM			
ASM_RIGHT	1	Datum Plane	
ASM_TOP	2	Datum Plane	
ASM_FRONT	3	Datum Plane	
TUT_BASE.PRT	4	Component	
ADTM1	5	Datum Plane	
TUT_WIDGET_HANDLE.	6	Component	
TUT_WASHER.PRT	7	Component	
TUT_WASHER.PRT	8	Component	
TUT_LONG_SHAFT.PRT	9	Component	
TUT_HEXNUT.PRT	10	Component	
RIGHT	1	Datum Plane	
TOP	2	Datum Plane	
FRONT	3	Datum Plane	
Protrusion id 39	4	Protrusion	Revolve
Cut id 99	5	Cut	Extrude
Cut id 154	6	Cut	Revolve
Hole id 2009	7	Hole	
Chamfer id 2122	8	Chamfer	Edge
Cut id 2037	9	Cut	Helical Sweep
Insert Here			
Pattern (BOLT_WASHER	11	Pattern	PATTERN
Insert Here			

Figure 28 Assembly model tree showing part features

to constrain each component in the assembly. We will look into this in depth in Lesson 8.

Also in a later lesson, we will see how we can create components in the assembly environment. This means that existing components can provide references for the creation of new components. Thus, you can create a new part to fit exactly to a previous part (for example, a shaft in a hole) by using existing geometry rather than manually transferring dimensions from one part to another.

There are just a couple more items we want to cover this lesson. Turn shading and color back on, if they aren't already. Close the Browser and Navigator windows if they are open.

Exploding an Assembly

Exploded views of assemblies are often helpful in showing how they are put together. If you have ever worked on a do-it-yourself project or kit, you have probably seen an exploded view in the plans. Getting this view in Pro/E is a snap. Select

> *View > Explode >*
> *Explode View*

The position of each component in the exploded assembly is determined by a default. However, the *explode position* of each component can be easily changed, as has been done here. These positions are stored with the assembly. Note also that while the subassembly is exploded from the main assembly, the components in the sub-assembly (in this case) are not exploded from each other. This indicates that the

Figure 29 The exploded assembly

explode state of various components can be individually set. It is also possible to add the explode lines to the view (if you have the appropriately licensed module). We will spend some more time with layout of the exploded view in a later lesson. For now, unexplode the assembly with

> *View > Explode > Unexplode View*

Opening Parts in an Assembly

When an assembly is being worked on, you will frequently want to open up an individual part by itself. This is easy to do. In the graphics window (Selection filter set to *Smart* or *Parts*), left click on the blue handle (part TUT_HANDLE). It will highlight in red. Now hold down the right mouse button and select *Open* from the pop-up menu. The part will open in a new window by itself. Pro/E can have many separate parts and assemblies (and drawings) loaded into memory at once - these are called *in session* - each in a separate resizable window. To switch to another window, for example, back to the assembly, select (in the pull-down menus at the top)

Window > TUT_WIDGET.ASM

Notice that the TUT_BASE part is also listed there, along with any other windows you may have created in this session.

This concludes our look at the structure and layout of a simple assembly until later in the tutorial. As always, you are encouraged to experiment further with the commands we have covered and see what you can discover on your own.

Obtaining Hard Copy

You may want to produce some hard copy documentation from time to time as you go through these lessons. There are several basic kinds of hard copy:

1. Plain text containing model, component, or feature information, feature lists, the model tree.
2. Formatted web documents.
3. Images of the graphics area of the screen.
4. Detailed engineering drawings.

For the first type, whenever you ask for this type of information, Pro/E generally creates a printable text file in the working directory. Look for files with the extension *txt, inf, lst, err,* or *dat*. These can be printed with any simple text editor or word processor.

Web-formatted documents are available for any page viewed in the Browser window (such as the BOM report). Both "Save" and "Print" shortcut buttons are on the Browser toolbar.

Images of the graphics area ("screen shots") can be obtained easily if you have the appropriate printer attached to your system. Generally, this means a Postscript compatible printer. Different printing methods may be required depending on whether the image is shaded or not. Sometimes the Windows Print Manager will take care of this for you. To obtain a screen shot of a line drawing (hidden line or wireframe image), issue the commands

File > Print

or use the **Print** shortcut button on the top toolbar. In the **Print** window that opens up, in the Destination field select the appropriate output device. Windows users will usually select MS Print Manager. You might check out the options available by clicking the *Configure* button. Then select *OK* (twice). It is not recommended that you do this with shaded images.

For shaded images, your best bet is to select

File > Save a Copy

and in the **Type** pull-down list, you can find a number of image formats (TIFF, JPG, EPS,...). The image file will be created in your working directory. These images can be very high resolution (up to 600 dpi and 24 bit color), but are also very large (think megabytes). PTC has a special program available for very high quality rendering of images suitable for glossy marketing purposes. This allows you to have multiple light sources, shadows, reflections, textures, and so on.

A quick and dirty way of getting hard copy of a shaded image in Windows is to capture the screen contents to the clipboard. Use ALT-PrtScrn to capture the active window. Then paste the clipboard image (a bitmap) into a graphics utility program or word processor. This image is stuck at the resolution of your screen, however.

If the methods above do not work, your system administrator may have some special instructions that need to be followed.

Obtaining hard copy of engineering drawings will be covered later in these lessons. For these, there are extra considerations of sheet size, drawing scale, pen width, and so on.

Leaving Pro/ENGINEER

When you want to quit Pro/E entirely, you can leave by using the ***Exit*** command in the **File** menu or the X at the top-right corner. If you accidentally do this, you can cancel the command.

Helpful Hint

Unlike many programs, Pro/E will not automatically save anything for you (like a regular timed backup on some systems). This applies both during operation and when you exit. If you leave the program without saving new work, it is basically gone! Anyone who says they have never lost work this way is probably lying!

Depending on how your system has been set up, Pro/E *may* prompt you to save your work when you exit[17]. This includes any parts, assemblies, drawings, and so on, that are currently *in session* (stored in memory). You will be prompted individually for each object in session. Reply with a *Y* or *N* to save each object in the current working directory. A middle click will accept whatever default is shown. If you are sure you have saved the most recent version of all objects in session, you don't need to do that again so press *Q* (for Quit).

This completes Lesson #1. You will no doubt be relieved to know that it is by far the longest in this series. There was much fundamental material to deal with, however. You are strongly

---

[17] You can include the option *prompt_on_exit* in the configuration file accessible using ***Tools > Options***. The default for this option is *No*. See the Appendix.

encouraged to experiment with any of the commands that have been presented in this lesson. Explore the other parts in the widget assembly, and experiment with the view controls. The only way to become proficient with Pro/E is to use it a lot!

In the next lesson we will create our first part using simple basic features (protrusion, cut, hole) and spend some time learning about the Intent Manager in Sketcher. The important concept of *design intent* will be introduced, with examples.

Questions for Review

Here are some questions you should be able to answer at this time, or that may provoke some thought and/or further study:

1.   What mouse buttons are used to pan, spin, and resize the object?
2.   What is the purpose of the datum planes?
3.   What are three ways to get on-line help?
4.   How can you get a shaded image of the part?
5.   How do you turn the datum plane visibility on and off?
6.   How many ways can you think of to modify the value of a dimension?
7.   What is the difference in operation between *View > Shade* and the "Shading" shortcut button?
8.   Where does your system first look for stored part files? What is this called?
9.   What is meant by *preselection*?
10. What is meant by the term *regeneration*?
11. What is the regeneration sequence? How can you determine it for a part?
12. On what menu can you set the datum display, color toggle, and display style simultaneously?
13. How many ways can you get to the default view orientation? What are they?
14. How do you change the working directory?
15. What is the spin center?
16. What is the correct toolbar button to execute the following functions:
     a.   toggle display of datum planes
     b.   start View Mode
     c.   regenerate
     d.   create a new named view
     e.   turn off the spin center
17. What is the function of the following toolbar buttons:

a)   b)   c)   d)   e)

18. What are the three primary goals in creating models?
19. Describe all the shortcuts and functions associated with the middle mouse button.
20. What is the difference between *view* and *display* controls?
21. Why do datum planes have two colors? What are they?

22. How large is a datum plane?
23. How do you turn off the display of datum tags?
24. What is meant by the *object/action* command style? Give an example. Compare this to the *action/object* style. Give an example of that, too.
25. What is the meaning of the following icons? Where do they appear?

26. Where is the explode command for assemblies?
27. Is it possible to *Hide* a solid feature?
28. What happens if you preselect a dimension and then right click? This is useful if the screen gets really crowded.

Exercises

1. Try out the model player with the widget assembly.
2. Open the part TUT_CUBE.PRT. Practice manipulating the display of the cube to show each numbered side facing the front of the screen, with the number the right way up. Do this as fast as you can. Hint: turn color and shading on (the cube is translucent, so you can see the numbers on the hidden faces). If your system cannot handle transparency, use hidden line display.
3. Obtain a printed hard copy of a hidden-line display of the base part.
4. Obtain a printed hard copy of the model tree for the base part showing the additional columns.
5. Obtain a hard copy of the Navigator window showing the model information for the part TUT_BASE.
6. Obtain a hard copy of the Navigator window showing the feature information for the BASE feature in the part TUT_BASE.
7. It is possible in the assembly **tut_widget.asm** to position the handle part perpendicular to the base. How can you do that? Obtain a printed hard copy of the assembly (no hidden lines) in this position.
8. Obtain a printed hard copy of a shaded image of the exploded view of the assembly **tut_widget.asm**. Label this with the part names. Bonus marks if you can figure out how to do this with *3D Notes*.
9. Make an isometric sketch of the three datum planes (TOP, FRONT, RIGHT). Identify two ways to select references for each of the standard engineering drawing views (top, front, right, left, back, bottom).

This page left blank.

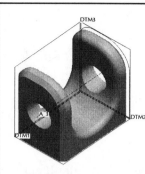

Lesson 2

Creating a Simple Object (Part I)
Introduction to Sketcher

Synopsis

Creating a part; introduction to Sketcher; Sketcher constraints; creating datum curves, protrusions, cuts; using the dashboard; saving a part; part templates.

Overview of this Lesson

The main objective this lesson is to introduce you to the general procedures for creating sketched features. We will go at quite a slow pace and the part will be quite simple (see Figure 1 on the next page), but the central ideas need to be elaborated and emphasized so that they are very clearly understood. Some of the material presented here is a repeat of the previous lesson - take this as an indication that it is important! Here's what we are going to cover:

1. Creating a Simple Part
2. Feature Types and Menus
3. Introduction to Sketcher
   ‣ Sketcher menus
   ‣ Intent Manager and Sketcher constraints
4. Creating a Datum Curve
5. Creating an Extruded Protrusion
   ‣ Using the Dashboard
6. Creating an Extruded Cut
7. Saving the part
8. Using Part Templates

It will be a good idea to browse ahead through each section to get a feel for the direction we are going, before you do the lesson in detail. There is a lot of material here which you probably won't be able to absorb with a single pass-through.

Start Pro/E as usual. If it is already up, close all windows (except the base window) and erase all objects in session using *File > Erase > Current* and *File > Erase > Not Displayed*. Close the Navigator and Browser windows.

Creating a Simple Part

In this lesson, we will create a simple block with a U-shaped central slot. By the end of the lesson your part should look like Figure 1 below. This doesn't seem like such a difficult part, but we are going to cover a few very important and fundamental concepts in some depth. Try not to go through this too fast, since the material is crucial to your understanding of how Pro/E works. We will be adding some additional features to this part in the next lesson.

Not only are we going to go slowly here, but we are going to turn off some of the default actions of Pro/E. This will require us to do some things manually instead of letting the program do them automatically. This should give you a better understanding of what the many default actions are. Furthermore, eventually you will come across situations where you don't want the defaults and you'll need to know your way around the program.

The first thing to do here is to close the Navigator and Browser panes.

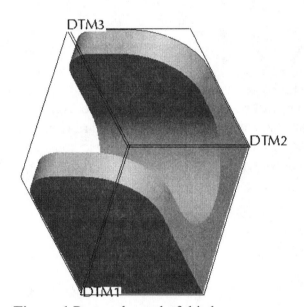

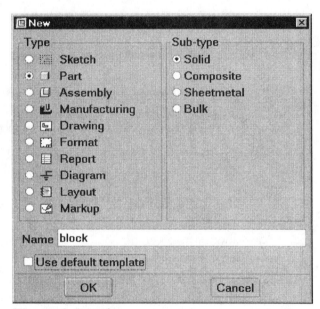

Figure 1 Part at the end of this lesson

Figure 2 Creating a new part

Creating and Naming the Part

Click the *Create New Object* short-cut button, or select *File > New*. A window will open (Figure 2) showing a list of different types and sub-types of objects to create (parts, assemblies, drawings, and so on). In this lesson we are going to make a single solid object called a *part*. Keep the default radio button settings

> *Part | Solid*

IMPORTANT: Turn off (remove the check) the *Use Default Template* option at the bottom. We will discuss templates at the end of this lesson.

Many parts, assemblies, drawings, etc. can be loaded simultaneously (given sufficient computer memory) in the current session. All objects are identified by unique names[1]. A default name for the new part is presented at the bottom of the window, something like **[PRT0001]**. It is almost always better to have a more descriptive name. So, double click (left mouse) on this text to highlight it and then type in

[block]

(without the square brackets) as your part name and press *Enter* or select *OK*.

The *New File Options* dialog window opens, as shown to the right. Since we elected (in the previous window) to not use the default template for this part, Pro/E is presenting a list of alternative templates defined for your system. As mentioned previously, we are going to avoid using defaults this time around. So, for now, select

Empty | OK.

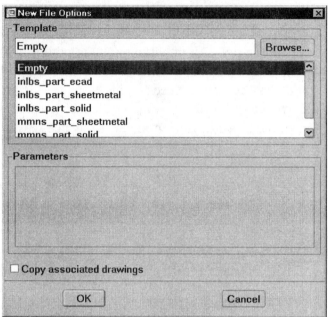

Figure 3 Setting options for new parts

At this time, **BLOCK** should appear in the title area of the graphics window. Also, some of the toolbar icons at the right are now "live" (ie. not grayed out).

Create Datum Planes and Coordinate System

We will now create the first features of the part: three reference planes to locate it in space. These are called *datum planes*. It is not absolutely necessary to have datum planes, but it is a very good practice, particularly if you are going to make a complex part or assembly. The three default datum planes are created using the "Datum Plane" button on the right toolbar, as shown in Figure 4. Note that these icons look quite similar to the buttons on the top toolbar that control the display of datums. What's the difference?

Select the *Datum Plane* button now.

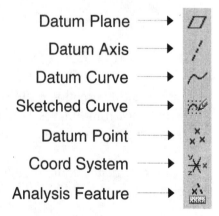

Figure 4 Right toolbar buttons for creating *DATUMS*

---

[1] Pro/E can keep track of objects of different types with the same names. For example a part and a drawing can have the same name since they are different object types.

The datum planes represent three orthogonal planes to be used as references for features to be created later. You can think of these planes as XY, YZ, XZ planes, although you generally aren't concerned with the X,Y,Z form or notation. Your screen should have the datum planes visible, as shown in Figure 5. (If not, check the datum display button in the top toolbar.) They will resemble something like a star due to the default 3D viewing direction. Note that each plane has an attached tag that gives its name: **DTM1**, **DTM2**, and **DTM3**. This view may be somewhat hard to visualize, so Figure 6 shows how the datum planes would look if they were solid plates in the same orientation. An important point to note is, while the plates in Figure 6 are finite in size, the datum planes actually extend off to infinity. Finally, before we move on to the next topic, notice that the last feature created (in this case DTM3), is highlighted in red. This is a normal occurrence and means that the last feature created is always preselected for you as the "object" part of the object/action command sequence.

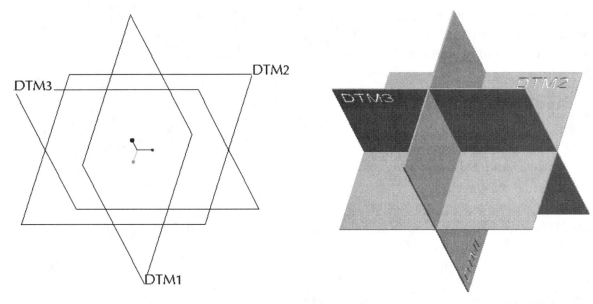

Figure 5 Default datum planes   **Figure 6** Datum planes as solid plates

Pro/ENGINEER Feature Overview

Below the datum creation buttons in the toolbar on the right are three other groups of buttons. These are shown in Figures 7, 8, and 9. If you move the cursor over the buttons, the tool tip box will show the button name.

Two of these menus contain buttons for creating features, organized into the following categories:

Placed Features (Figure 7) - (holes, rounds, shells, ...) These are features that are created directly on existing geometry. Examples are placing a hole on an existing surface, or creating a round on an existing edge of a part.

Sketched Features (Figure 8) - (extrusions, revolves, sweeps, blends, ..) These features require the definition of a two-dimensional cross section which is then manipulated into the third dimension. Although they usually use existing geometry for references, they do not specifically require this. These features will involve the use of an important tool called Sketcher.

The final group of buttons (Figure 9) is used for editing and modifying existing features (copy, mirror, pattern, ...). We will deal with some of these commands later in the Tutorial.

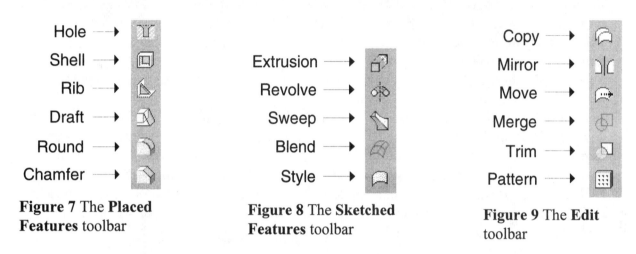

Figure 7 The **Placed Features** toolbar

Figure 8 The **Sketched Features** toolbar

Figure 9 The **Edit** toolbar

In this lesson we will be using the ***Extrude*** command to create two types of sketched features (a protrusion and a cut). In the next lesson, we will use the ***Hole***, ***Round***, and ***Chamfer*** commands to create three placed features. Before we continue, though, we must find out about an important tool - Sketcher.

Introducing Sketcher

Sketcher is the single, most important tool for creating features in Pro/E. It is therefore critical that you have a good understanding of how it works. We will take a few minutes here to describe its basic operation and will explore the Sketcher tools continually through the next few lessons. It will take you a lot of practice and experience to fully appreciate all that it can do.

Basically, Sketcher is a tool for creating two-dimensional figures. These can be either stand-alone features (*Datum Curves*) or form the cross sectional shape of some solid features. The aspects of these figures that must be defined are shape, location, and size. Within Sketcher you will find the usual (and expected) drawing tools for lines, arcs, circles, and so on, to create the shape. The location and size aspects are handled by specifying alignments with or dimensions to existing geometry.

Sketcher is really "smart", that is, it will anticipate what you are going to do (usually correctly!) and do many things automatically. Occasionally, it does make a mistake in guessing what you want. So, learning how to use Sketcher effectively involves understanding exactly what it is doing for you (and why) and discovering ways that you can easily over-ride this when necessary.

The "brain" of Sketcher is called the *Intent Manager*. We will be discussing the notion of *design intent* many times in this tutorial. In Sketcher, design intent is manifest not only in the shape of the sketch but also in how constraints and dimensions are applied to the drawing so that it is both complete and conveys the important design goals for the part. Completeness of a drawing implies that it contains just enough geometric specification so that it is uniquely determined. Too little information means the drawing is under-specified; too much means that it is over-specified. The function of Intent Manager is to make sure that the sketch always contains just the right amount of information. Moreover, it tries to do this in ways that, most of the time, make sense. Much of the frustration involved in using Sketcher arises from not understanding (or even sometimes realizing) the nature of the choices it is making for you or knowing how easy it is to make alternate choices. When you are using Sketcher, Intent Manager must be treated like a partner - the more you understand how it works, the better the two of you will be able to function[2].

The term *sketch* comes from the fact that you do not have to be particularly exact when you are "drawing" the shape, as shown in the two figures below. Sketcher (or rather Intent Manager) will interpret what you are drawing within a built-in set of rules. Thus, if you sketch a line that is approximately vertical, Sketcher assumes that you want it vertical. If you sketch two circles or arcs that have approximately the same radius, Sketcher assumes that's what you want. In cases like this, you will see the sketched entity "snap" to a particular orientation or size as Intent Manager fires one of the internal rules.

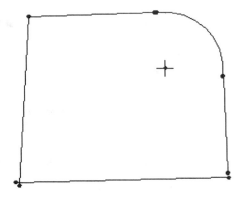

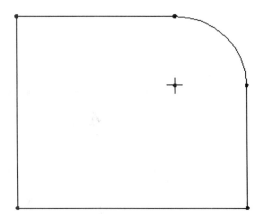

Figure 10 Geometry input by user. Note misaligned vertices, non-parallel edges, non-tangent curves.

Figure 11 Geometry after processing by Sketcher. Note aligned vertices, parallel edges, tangent curves.

When Sketcher fires one of its internal rules (this occurs while you are sketching), you will be alerted by a symbol on the sketch that indicates the nature of the assumed condition. If you accept the condition, it becomes a *constraint* on the sketch. These symbols are summarized in Table 2-1 below. You should become familiar with these rules or constraints, and learn how to use them to your advantage. Conversely, if you do not want a rule invoked, you must either

---

[2] Intent Manager was introduced several releases ago. Some veteran Pro/E users still have not made the switch from "the old days". For those users, Pro/E has the ability to turn off the Intent Manager and let them do everything manually. This tutorial will not discuss the use of Sketcher in this old style - it really isn't as efficient as Intent Manager.

(a) use explicit dimensions or alignments, or

(b) exaggerate the geometry so that if fired, the rule will fail, or

(c) tell Pro/E explicitly to ignore the rule (disable the constraint).

You will most often use option (a) by specifying your desired alignments and dimensions and letting Sketcher worry about whatever else it needs to solve the sketch. When geometry is driven by an explicit dimension, fewer internal rules will fire. Option (b) is slightly less common. An example is if a line in a sketch must be 2° away from vertical, you would draw it some much larger angle (like 15° or so) and put an explicit dimension on the angle. This prevents the "vertical" rule from firing. Once the sketch has been completed with the exaggerated angle, you can modify the dimension value to the desired 2°. For method (c), there is a command available that explicitly turns off the rule checking (for all rules or selected ones only) during sketching. This is very rarely used.

Table 2-1 Implicit Constraints in Sketcher

Rule	Symbol	Description
Equal radius and diameter	R	If you sketch two or more arcs or circles with approximately the same radius, the system may assume that the radii are equal
Symmetry	→ ←	Two vertices may be assumed to be symmetric about a centerline
Horizontal or vertical lines	H or V	Lines that are approximately horizontal or vertical may be considered to be exactly so.
Parallel or perpendicular lines	∥ or ⊥	Lines that are sketched approximately parallel or perpendicular may be considered to be exactly so.
Tangency	T	Entities sketched approximately tangent to each other may be assumed to be tangent
Equal segment lengths	L	Lines of approximately the same length may be assumed to have the same length
Point entities lying on other entities or collinear with other entities	—○—	Point entities that lie near lines, arcs, or circles may be considered to be exactly on them. Points that are near the extension of a line may be assumed to lie on it.
Equal coordinates	▪ ▪	Endpoints and centers of the arcs may be assumed to have the same X- or the same Y-coordinates
Midpoint of line	M	If the midpoint of a line is close to a sketch reference, it will be placed on the reference.

An example of a sketch with the geometric constraints is shown in Figure 14. Note how few dimensions are required to define this sketch. See if you can pick out the following constraints:

- ▸ vertical lines
- ▸ horizontal lines
- ▸ perpendicular lines
- ▸ tangency
- ▸ three sets of equal length lines
- ▸ equal radius
- ▸ vertical alignment (two cases)

How do you suppose Sketcher is able to determine the radius of the rounded corners (fillets) at the top and bottom on the left edge?

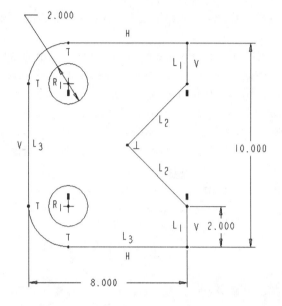

Figure 14 A regenerated sketch showing implicit constraints

In this lesson, we will use two methods to create a couple of sketched features. The two methods differ in how they use sketcher to define the cross section of the features. The method you use in your own modeling is a matter of personal preference. Both features we will make are extrusions: one will be a *protrusion* (which adds material) and the other is a *cut* (which removes material). Either of the two methods shown here can be used to create either protrusions or cuts; for either method, whether you add or remove material is determined by a single mouse click!

In the first method, we invoke Sketcher first to create the cross sectional shape of the extrusion. This shape is defined in a *sketched datum curve* which becomes a stand-alone feature in the model. We then launch the extrude command, specifying the datum curve to define the cross section of the feature. In the second method, we do not create a separate curve but rather invoke Sketcher from inside the extrusion creation sequence.

Creating a Sketched Curve

When we left the model last, the datum plane DTM3 was highlighted in red. If that is not the case now, use preselection highlighting to select that datum now.

In the datum toolbar on the right of the screen, pick the **Sketched Datum Curve** button. Be careful not to pick the button just above it - that one will create a datum curve using sets of existing datum points, points read from a file, or using equations. If you accidentally pick the wrong button, you can back out with the **Quit** command.

Setting Sketch Orientation

The **Sketched Datum Curve** dialog window opens as shown in Figure 15. Since DTM3 was highlighted (in red) prior to the present command, it has been preselected as the Sketch Plane. It is now highlighted in the graphics window in orange. This is the plane on which we will draw the sketch. Notice the yellow arrow attached to the edge of DTM3 pointing back into the screen. This is the direction of view onto the sketch plane. You can reverse that with the **Flip** button in the dialog window. Leave it pointing towards the back. DTM1 is now highlighted in red in the graphics window. In the dialog window, DTM1 is identified as the Sketch Orientation **Reference**, with the **Orientation** set to **Right**. What is all this about?

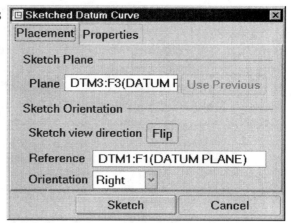

Figure 15 Defining the sketch plane and orientation for a *Sketched Curve*

The relation between the sketch plane and the sketch orientation reference generally causes a lot of confusion for new users, so pay attention!

The meaning of the sketch plane is pretty obvious - it is the plane on which we will draw the sketch - in this case DTM3. Our view is always perpendicular to the sketch plane[3]. That is not enough by itself to define our view of the sketch since we can be looking at that plane from an infinite number of directions (imagine the sketch plane rotating around an axis perpendicular to the screen). The **Orientation** option list in the dialog window (**Top**, **Bottom**, **Left**, **Right**) refers to directions relative to the computer screen, as in "TOP edge of the screen" or "BOTTOM edge of the screen" and so on. We must combine this orientation with a chosen reference plane (*which must be perpendicular to the sketch plane*) so that we get the desired direction of view onto the sketching plane.

In the present case, when we get into Sketcher we will be looking directly at the brown (positive) side of DTM3. So that the sketch is the right way up, we can choose either DTM2 to face the Top of the screen, or (as was chosen automatically for us) DTM1 can face the Right of the screen. Note that both DTM1 and DTM2 are both perpendicular to the sketch plane, as required. The direction a plane or surface "faces" is determined by its normal vector. The normal vector for a datum plane is perpendicular to the brown side. For a solid surface, the orientation is determined by the outward normal.

Read the last couple of paragraphs again, since new users are quite liable to end up drawing their sketches upside-down!

---

[3]Well, almost always. It is possible to sketch in 3D, in which case you can manipulate your view so that you are not looking perpendicularly at the sketch plane. We will not attempt that here.

To illustrate the crucial importance of the reference plane, consider the images shown in Figure 16. These show two cases where the same sketching plane **DTM3** was used, the same sketched shape was drawn, the same reference orientation TOP was chosen, but where different datums were chosen as the sketching reference. On the left, the TOP reference was **DTM2**. On the right, the TOP reference was **DTM1**. The identical sketch, shown in the center, was used for both cases (rounded end of sketch towards the top of the screen). However, notice the difference in the orientation of the part obtained in the final shaded images. Both of these models are displayed in the default orientation (check the datum planes). Clearly, choosing the sketching reference is important, particularly for the base feature.

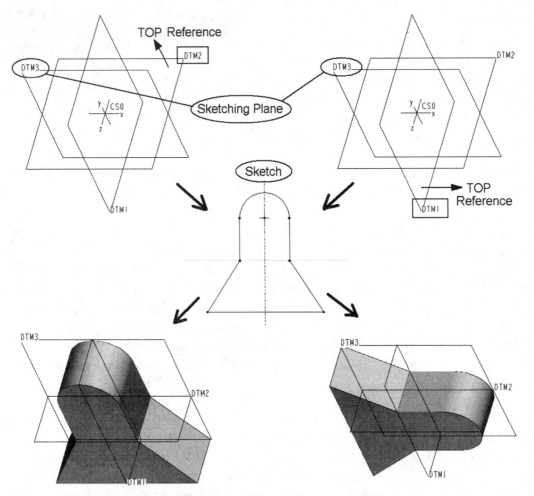

Figure 16 The importance of the sketching reference plane!

Let's continue on with creating the curve. Make sure the **Sketched Datum Curve** dialog window is completed as in Figure 15. Select the *Sketch* button in the dialog window.

Several things will happen: the graphics window color changes to black, two dashed gray lines appear that cross in center of the screen, an orange square appears to indicate our sketching plane (DTM3).

At the top right of the screen is another **References** dialog window. In this window we select existing geometry to help Sketcher locate the new sketch relative to the part. In the present case, there isn't much to choose from, and two references have been chosen for us - DTM1 and DTM2. These references are responsible for the two dashed lines in the graphics window. The number of references you choose is not limited - there may be several listed here. You are also free to delete the ones chosen for you. However, notice the **Reference Status** at the bottom of this dialog. **Fully Placed** means enough references have been specified to allow Sketcher to locate your sketch in the model. If there are not enough references, the status will be **Partially Placed**. For now, do not proceed beyond

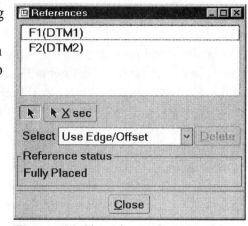

Figure 17 Choosing references in Sketcher

this window unless you have a **Fully Placed** status indicated. Once you have that, select *Close* in the References window.

The drawing window is shown in Figure 18. Note that you are looking edge-on to the datums DTM1 and DTM2. The datum DTM1 (actually, its brown side) is facing the right edge of the screen, as specified in the dialog back in Figure 15. Note that we could have obtained the same orientation by selecting DTM2 to face the top of the screen.

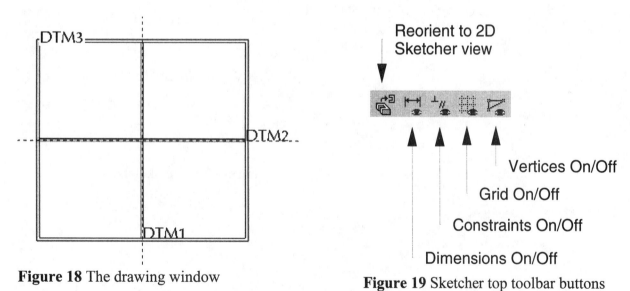

Figure 18 The drawing window

Figure 19 Sketcher top toolbar buttons

Another change is the addition of some new toolbar buttons at the top of the screen. These control the display of entities in the sketch. See Figure 19. The four buttons with the eyeballs control display of dimensions, constraints, the grid (default off), and vertices. Leave these buttons in their default position. It is seldom (if ever) that you will need to turn on the grid in Sketcher. The button at the left will return you to the default view of the sketch if you should accidentally (or intentionally) go into 3D view.

The Sketcher Toolbar

The major addition to the screen is the new toolbar on the right of the screen. This contains the Sketcher tools and is shown in Figure 20. Several buttons on this menu have fly-outs, indicated by the ➤ symbol on the right edge. These fly-outs lead to related buttons, and are listed in Table 2-2. Compared to some 2D drawing programs, this doesn't seem like such a large number of drawing commands. Rest assured that there will not be much that you cannot draw with these[4].

When you are sketching, many of the commands in the right toolbar are instantly available (but context sensitive) by holding down the right mouse button in the graphics window. This will bring up a pop-up window of commands relevant in the current situation.

A handy hint: from wherever you are in the Sketcher menu structure, a single middle mouse click will often abort the current command and return you to the toolbar with the *Select* command already chosen. Sometimes, you may have to click the middle button twice.

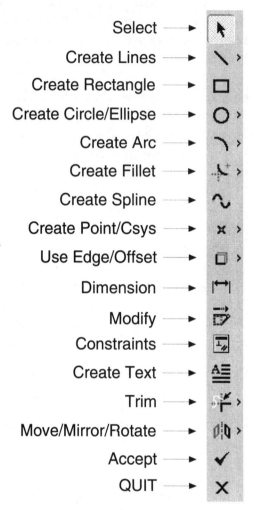

Figure 20 The Sketcher toolbar

---

[4]One command that some people miss is for creating regular polygons - like a hexagon. Once you get used to the Sketcher commands, though, even that is easy to do with the existing drawing tools.

Table 2-2 Sketcher Toolbar Flyout Buttons

Button Flyout Group	Button Commands
![flyout]	Line | Tan-Tan Line | Centerline
![flyout]	Circle | Concentric | 3 Point | 3 Tan | Ellipse
![flyout]	Tangent End | Concentric | Center | 3 Tan | Conic Arc
![flyout]	Circular fillet | Conic fillet
![flyout]	Point | Coordinate System
![flyout]	Use Edge | Offset edge
![flyout]	Dynamic trim (delete) | Trim(extend) | Divide
![flyout]	Mirror | Rotate | Move

Creating the Sketch

Select the *Line* tool using one of the following three methods:

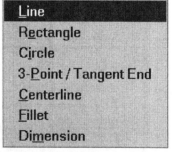

Figure 21 RMB pop-up menu in Sketcher

- using the *Line* toolbar button on the right, OR
- in the pull-down menus select *Sketch > Line > Line*, OR
- hold down the right mouse button and select *Line* from the pop-up menu (Figure 21).

You will now see a small yellow X which will chase the cursor around the screen. Notice that the X will snap to the dashed references when the cursor is brought nearby. While you are creating the figure, watch for red symbols (V, H, L) that indicate Intent Manager is firing an internal rule to set up a constraint (Vertical, Horizontal, Equal Length). These symbols will come and go while you are sketching. The trick with Sketcher is to get Intent Manager to fire the rule you want, then click the left mouse button to accept the position of the vertex. Click the corners in the following order. After each click, you will see a straight line rubber-band from the previous position to the cursor position:

1.   left-click at the origin (intersection of **DTM1** and **DTM2**)
2.   left-click above the origin on **DTM1** (watch for V)
3.   left-click horizontally to the right (watch for H and L - we do not want L)
4.   left-click straight down on **DTM2** (watch for V)
5.   left-click back at the origin (watch for H)
6.   middle-click anywhere on the screen

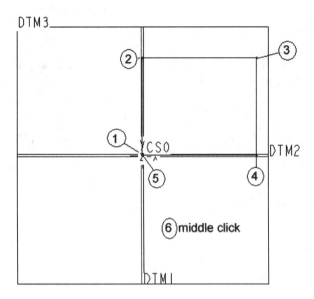

Figure 22 Drawing the Sketch

When you are finished this sequence, you are still in **Line** creation mode (notice the yellow X on screen and the *Line* toolbar button). If you middle click again, you will leave that and return to *Select* mode - the same as if you picked on the *Select* button in the right toolbar, but much faster.

The sketched entities are shown in yellow. Note that we didn't need to specify any drawing coordinates for the rectangle, nor, for that matter, are any coordinate values displayed anywhere on the screen. This is a significant departure from standard CAD programs. We also didn't need the grid or a grid snap function (although both of these are available in Pro/E).

You can also sketch beyond the displayed edges of the datum planes - these actually extend off to infinity. The displayed extent of datum planes will (eventually) adjust to the currently displayed object(s).

Helpful Hint

If you make a mistake in drawing your shape, here are some ways to delete entities:

1. Pick the *Select* tool in the right toolbar and left click on any entity you want to delete. Then either press the *Delete* key on the keyboard, or hold down the RMB and choose *Delete*.

2. If there are several entities to delete, hold the CTRL key down while you left click on each entity. Then pick *Delete* as before.

3. You can click and drag to form a rectangle around a set of entities. Anything completely inside the rectangle is selected. Use *Delete* as before.

4. Notice the *Undo* and *Redo* buttons on the top toolbar

We will cover more advanced Sketcher commands for deleting and trimming lines a bit later.

After you have finished the sequence above, Sketcher will put two dimensions on the sketch - for the height and width of the rectangle. These will be in gray, so may be hard to see, but similar to those shown in Figure 23. For the first feature in a part, the numerical values of these dimensions are picked more-or-less at random (although they are in correct proportion to each other). For later features in the part, Sketcher will know the sketch size more accurately because it will have some existing geometry to set the scale.

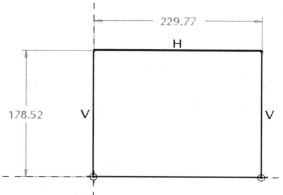

Figure 23 Completed sketch with weak dimensions

Weak vs Strong Dimensions

A dimension created by Sketcher is called "weak" and is shown in gray. Strong dimensions, on the other hand, are those that you create. You can make a strong dimension in any of three ways:

    ☞   modify the value of a weak dimension

or ☞   create a dimension from scratch by identifying entities in the sketch and placing a new dimension on the drawing

or ☞   select a weak dimension and promote it to strong using the RMB pop-up menu

Strong dimensions will be shown in yellow.

The special significance of weak and strong dimensions is as follows. When Intent Manager is "solving" a sketch, it considers the sketch references, any implicit rules that have fired (like H, V, and so on) and any existing dimensions. If there is not enough information to define the drawing, Sketcher will create the necessary and sufficient missing dimensions. These are the weak dimensions. If Sketcher finds the drawing is overconstrained (too many dimensions or constraints) it will first try to solve the sketch by deleting one or more of the weak dimensions (the ones it made itself earlier). It will do this without asking you. This is one way for you to override Intent Manager - if you don't like the dimensioning scheme chosen by Sketcher, just create your own (automatically strong) dimensions. Sketcher will remove whichever of the weak dimensions are no longer needed to define the sketch. Sketcher assumes that any strong dimensions you have created shouldn't be messed with! However, if Sketcher still finds the drawing overconstrained, it will tell you what the redundant information is (which may be dimensions or constraints), and you can choose what you want deleted. Thus, although weak dimensions can be deleted without asking you, Sketcher will never delete a strong dimension without your explicit confirmation.

We want to modify the two weak dimensions on the rectangle in a couple of ways. First, we can make a cosmetic improvement by selecting the dimension text (the number) and performing a drag-and-drop to move it to a better location. Note in passing that preselection highlighting also works with dimensions and constraints.

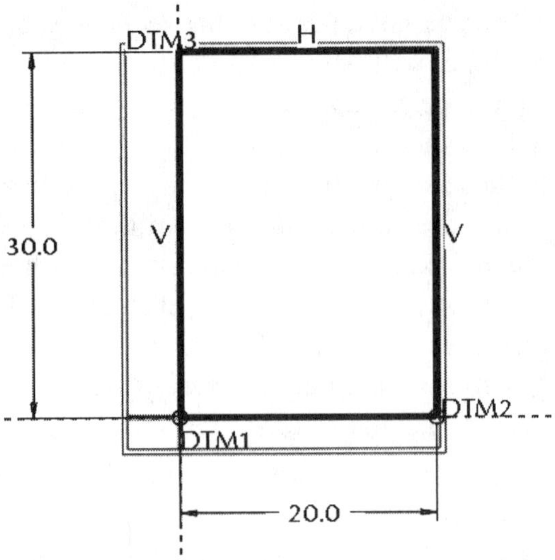

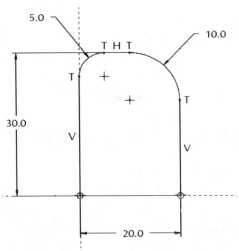

Figure 24 Modified sketch

Next we want to change the numeric value of the dimension. Double-click on the horizontal dimension. In the text entry box, enter the value **20**. When you hit Enter, the sketch geometry will be updated with this new dimension. The dimension is now strong. Change the vertical dimension to **30**. It will also now be strong. (Click anywhere on the graphics window to remove the red highlight.) See Figure 24. Notice that the indicated extent of the datum plane DTM3 adjusts to the sketch.

Now we'll add a couple of rounded corners, technically known as fillets, on the top corners of the sketch to help us "see" the orientation of the feature in 3D. Select the *Fillet* toolbar button on the right (or from the RMB pop-up menu) and pick on the top and right lines in the sketch close to but not at the corner. A circular fillet is created to the pick point closest to the corner. Two tangent constraints (T) are added, along with a weak dimension for the fillet radius. Do the same on the top and left lines. Middle click to return to Select mode. Because our fillet command has removed two vertices on the top of the sketch, Intent Manager has removed our two strong dimensions (which used those vertices) and replaced them with weak ones. You can make them strong by selecting them, clicking the RMB, and selecting *Strong* in the pop-up menu. Modify the cosmetics and values of the fillet radius dimensions as shown in Figure 25.

Figure 25 Completed sketch

This completes the creation of our sketched datum curve. Select the *Accept* (or *Continue*) toolbar icon. This returns us to the regular graphics window with our new sketched curve shown in red (last feature created). You can spin the model around with the middle mouse button to see this curve from different view points. When you are finished with this, return the model to approximately the default orientation - Figure 26.

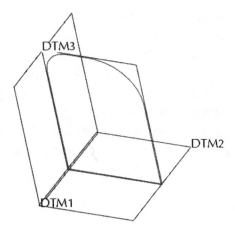

Figure 26 Datum curve - default orientation

Creating a Solid Protrusion

Most of the work to create this protrusion has been done already - creating the sketched datum curve that defines its shape. This curve should be highlighted in red. If you have been playing around with the model and the sketch is blue, just left click on it to select it again.

There are a number of ways to launch the protrusion creation command. With the sketched datum curve highlighted, the easiest way is to pick the *Extrude* button in the right toolbar.

A message window will open (**Section Selection**). Read this carefully. The essence of the message is that the sketched datum curve is going to be copied into the extrude feature, but once the feature is created the datum curve will not be associated with it. We will find out that this behavior in Pro/E is somewhat unusual - hence the message window. Normally, when a feature is used as a reference to create another feature, there is a parent/child relation set up. In essence, what is going to happen here is that Pro/E will make a copy of the sketched datum curve in the extrude feature and no parent relation will be created to the curve. The message window gives you a chance to turn off future appearances of itself. You should probably not do that - just press *Continue*.

What you will see now is a yellow shaded image of the protrusion, Figure 27. On this shape, you will see a yellow arrow that indicates the extrusion direction, which by default comes off the positive side of the sketch. There is also a dashed line ending in a white square. This is a drag handle. Click on this with the mouse and you can drag it to change the length of the extrusion. This length is also shown in a dimension symbol. You can even drag this out the back of the sketch to extrude in the opposite direction. This direct manipulation of the feature on the screen is called, in Pro/E vernacular, *Direct Modeling*. Bring the protrusion out the front and double click on the numeric dimension, and enter the value **30**.

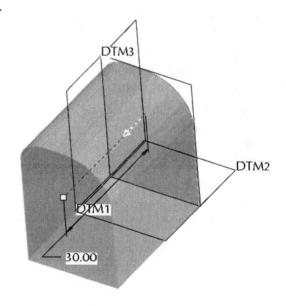

Figure 27 The protrusion preview

At the bottom of the graphics window is a new collection of tools. These comprise the *Dashboard*, which is a major innovation in the Pro/E Wildfire interface. Many features are constructed with tools arranged using this new interface element. It is worth spending some time exploring this one in detail, since you will probably be using it the most.

The Extrude Dashboard

The dashboard collects all of the commands and options for feature creation in an easily navigated interface. Moreover, most optional settings have been set to default values which will work in the majority of cases. You can change options at any time and in any order. This is a welcome and significant departure from previous releases of Pro/E.

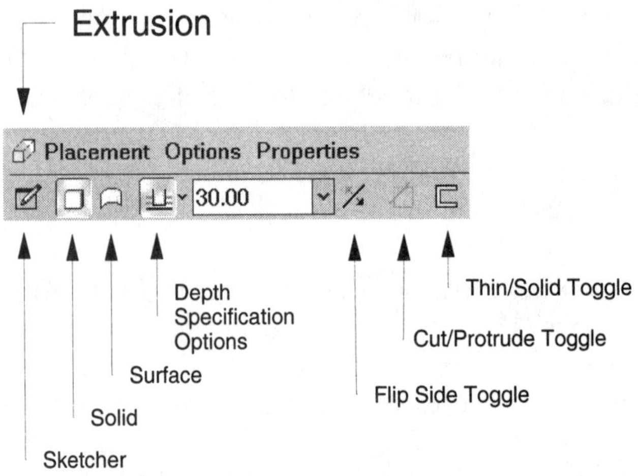

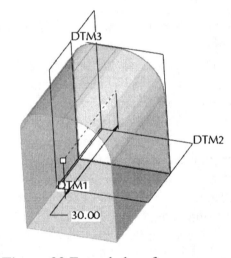

Figure 28 The *Extrude* dashboard

The dashboard contains two areas. On the left (Figure 28) are commands, settings, and so on for the particular feature under construction. On the top row, the feature is identified with the toolbar icon - extrusion in this case. The icons on the second row operate as follows:

Sketcher - selecting this will launch you into Sketcher to select a sketch plane, sketch reference, and so on. We don't need to go there now because we already have a sketch. If you wanted to change the sketch, this is how you access it. You can also enter this way to create a sketch for a new feature - we will go this route in the next feature.

Solid and *Surface* buttons - these are an either/or set. The default button is to create a solid. If you pick the next button, *Surface*, the sketch will be extruded as an infinitely thin surface (Figure 29). Return this to the *Solid* selection.

Figure 29 Extruded surface

Depth Spec Options - the next button is a pull-up menu that lists all the possibilities for setting the depth of the extrusion. These are indicated in Figure 30. The default is a *Blind* extrusion, which means the extrusion is for a fixed distance. Other options may appear here as more part geometry appears (as in the cut which we will do next).

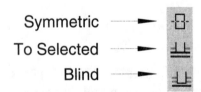

Figure 30 Depth Spec options

Blind Depth - this contains the numeric value of the length of the protrusion. If the depth specification on the button to the left is not Blind, this text input area is grayed out.

Flip - selecting this will reverse the direction of the protrusion (the yellow arrow).

Cut/Protrude - this allows you to change the meaning of the solid feature from a protrusion (which adds solid material) to a cut (which removes solid material). Since there is nothing to remove at this time, this command is grayed out - all we can do is add.

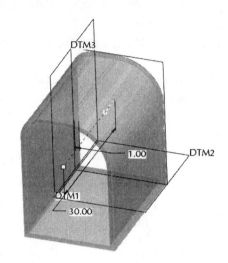

Solid/Thin - press this to see the solid block replaced by a thin-walled extrusion. A new dimension appears in the graphics window and on the dashboard. This is the thickness of the solid wall. Try changing this thickness to something like 1.0. On which side of the sketched curve has this been added? Another ***Flip*** button has also appeared. Press this a couple of times - it controls which side of the sketch the material is added to. Actually, it is a three way switch since you can also add material equally on both sides. Press the ***Solid/Thin*** button again to return to a full solid protrusion.

Figure 31 A ***Thin*** extruded solid

The slide-up panels do the following:

Placement - allows you to identify the section, or re-enter Sketcher

Options - more information about the depth specification. We will find out what is meant by "Side 2" in a later lesson.

Properties - specify the name of the feature

As you explore the creation of new features in Pro/E you should investigate what is in each of these menus. They are context sensitive, so there is a lot of variety in what you will find.

On the right end of the dashboard are several common tools that appear for all features. See Figure 32. These function as follows:

Pause - allow you to temporarily suspend work on this feature so that you can, for example, create a missing reference like a datum plane, measure something in the model, etc. When you are finished with the side trip, press the symbol ▸ that appears here to continue where you left off.

Preview - (default on = checked) this is responsible for the shaded yellow display of the feature under construction. Uncheck this - all you will see is the feature creation direction, drag handle, and depth dimension. Turn this back on.

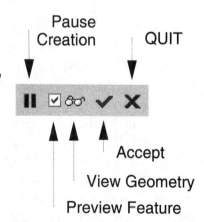

Figure 32 Common dashboard controls

View Geometry (or Verify) - this shows what the geometry will look like when the feature is fully integrated into the part. Not much happens with this first protrusion. Press again to return to preview.

Accept and *Quit* - do just what you expect!

Select *Accept*. The message window informs you that the feature has been created successfully. The block now appears, Figure 33, with its edges highlighted in red (last feature created).

We will now add another extruded feature - this time we will create a cut that removes material. Furthermore, instead of creating the sketched shape first, as we did for the solid protrusion, we will create the sketch within the feature itself.

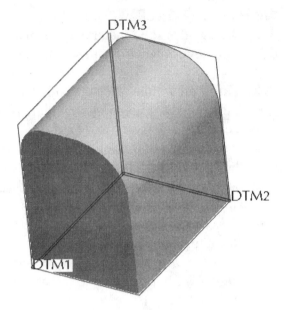

Figure 33 Completed protrusion

Creating an Extruded Cut

Start by launching the *Extrude* command from the right toolbar. The extrude dashboard at the bottom of the screen opens. Select the *Sketcher* button at the left end of the dashboard. The **Section** dialog window appears. This time, however, nothing has been preselected for us as it was for the previous sketch. We'll have to enter the data ourselves.

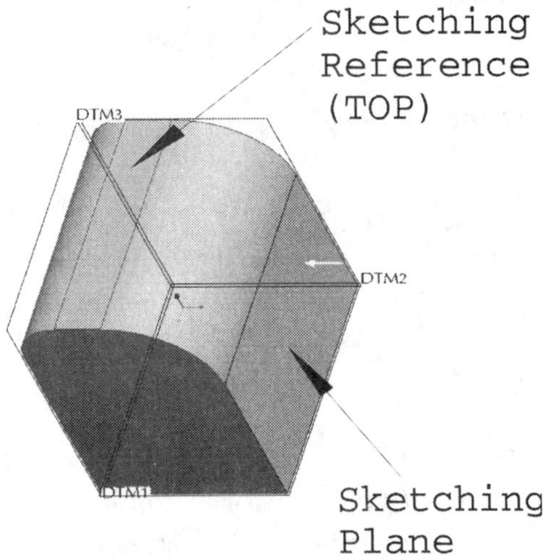

First, the dialog is waiting for you to select the sketching plane, so pick on the right side surface of the block (see Figure 34). Pay attention to preselection here. Notice the preselection filter setting; you will not be able to pick an edge or a curved surface (both of these would be illegal). As soon as you pick the sketching plane (it highlights in orange), a yellow arrow will appear showing our direction of view relative to the surface. The *Flip* button can be used to reverse this direction, but leave it as it is. Pro/E makes a guess at a potential reference plane for you to use. This may depend on the current orientation of your view, and might result in a strange view orientation in sketcher (like sideways or even upside down). We want to be a bit more careful and specific here. Pick on the top planar surface (Figure 34), between the two tangent lines of the rounded corners; the surface will highlight in red. In the *Orientation* pull-down list, select *Top* so that the reference will face the top of the screen. We now have our sketch plane and reference set up, so select *Sketch* at the bottom of the dialog window.

Figure 34 Setting up to sketch the cut

We are now in Sketcher (Figure 35). Two references have been chosen for us (the back and top surfaces of the object). These will be OK for now, so select *Close* in the **References** window. We are going to create the U-shaped figure shown in Figure 36. Note that there is no sketched line across the top of the U - there is no inside or outside. Thus, it is technically called an *open* sketch (as opposed to a *closed* sketch for our previous feature). There are some restrictions on the use of open sketches which we will run across in a minute or two. In general, try to keep your sketches closed - you will have fewer problems that way.

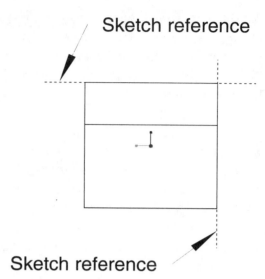

Figure 35 References for cut sketch

Use the RMB pop-up menu to select the *Line* command. Start your sketch at vertex 1 in Figure 36 - the cursor will snap to the reference. Then drag the mouse down and pick vertex 2 (note the V constraint), and middle click to end the *Line* command. Some weak dimensions will appear. Do nothing about them yet because, since they are weak, they are liable to disappear anyway. If we make them strong, this will cause us extra work dealing with Intent Manager. Wait until the geometry of the sketch is finished before you start worrying about the dimensioning scheme.

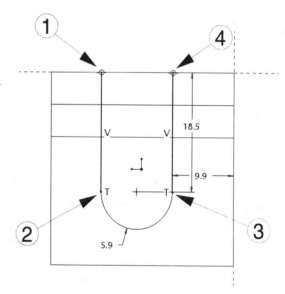

Figure 36 Sketch geometry

Use the RMB pop-up menu again and select the *3-Point/Tangent End* command. Pick on the end of the sketched line and drag the mouse downwards in the direction of tangency. Once the arc has been established, drag the cursor over to the right (the arc will rubber-band while maintaining the tangency constraint) and click at vertex 3. (If you drag straight across to vertex 3 you will get a 3-point arc which is not automatically tangent at vertex 2.) You should see two small blips that indicate when vertex 3 is at the same height as the center of the arc. Use the RMB menu to pick *Line* again.

Now left click at vertex 3 and draw a vertical line up to snap to the reference at vertex 4. Our sketch is complete. Use the middle mouse button to return to Select mode. Your dimensions may be different from those shown in Figure 36. Your dimensioning scheme may even be slightly different. It may be easier to see this if you go to hidden line display instead of shading.

All the dimensions should be weak. Drag them to a better location if necessary (off the part). Compare the dimensioning scheme with the one in Figure 37. We want to have a dimension from the reference at the back of the part to the center of the arc of the U. If you do not have that dimension, we'll have to add one manually. This will illustrate a case where we will override the Intent Manager.

To create your own dimension, select the *Dimension* command from the right toolbar. Click on the vertex at the center of the arc (it will highlight) then click again on the dashed reference line at the right. Now middle click in the space above the part where you want the dimension text to appear. It's that easy! Note that this dimension shows immediately in yellow since it is strong. One of the weak linear dimensions should be gone. Middle click to get back to *Select*.

Modify the values of the dimensions to match those in Figure 37.

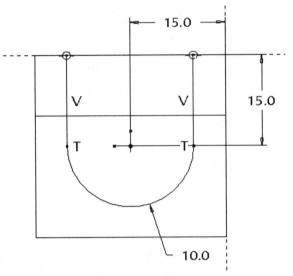

Figure 37 Final sketch for cut

The sketch is now complete, so click on the *Accept* button in the toolbar. If you are in hidden line display, return to shading display.

The feature will now be previewed. A couple of new buttons have appeared on the dashboard. First, in the **Depth Spec** pull-up list, there are a few more options available (Figure 38). For this cut, we would like the sketch to be extruded through the entire part, so pick the *Through All* option. Note that the dimension for a blind extrusion disappears from the screen. To the right of this area, click the *Flip* button to make the extrusion go through the part. The *Remove Material* button needs to be selected. Now, there are two yellow arrows attached to the sketch. The one perpendicular to the plane of the sketch shows the direction of the extrusion. The other shows which side of the sketched line we want to remove material from. These should be set as shown in Figure 39.

Blind ——▶
Symmetric ——▶
To Next ——▶
Through All ——▶
Through Until ——▶
To Selected ——▶

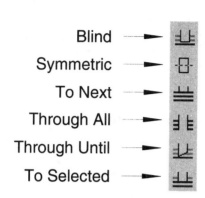

Figure 38 More Depth Spec options in the dashboard

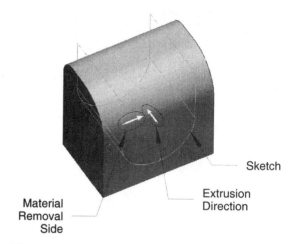

Sketch

Extrusion Direction

Material Removal Side

Figure 39 Defining cut attributes (direction and material removal side)

Now select the *Verify* button in the right area of the dashboard. If you have the ***Remove Material*** button set wrong, that is for a protrusion instead of a cut, Pro/E will not be able to create the feature. It will display thick dark green lines on the block and give you an error message about "unattached protrusion". The source of this problem is the open sketch for the U. This sketch is ambiguous since when the sketch starts out from the sketch plane, the vertices at the ends are out in the open air; Pro/E does not know how to create the solid to attach it to the existing part. This problem does not occur with a cut as long as the open ends of the sketch stay outside or on the surface of the part.

Another common error with cuts is having the material removal side set wrong (the second yellow arrow in Figure 39). If you do that for this part, you will end up with Figure 40. Make sure the material removal arrow points to the inside of the U. Plus, you should explore the **Placement**, **Options**, and **Properties** menus on the dashboard before you leave.

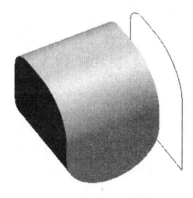

Figure 40 Removing from the wrong side of the sketch

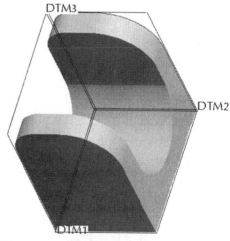

DTM3

DTM2

DTM1

Figure 41 Cut feature completed

We are finished creating this feature, so select the *Accept* button at the right end of the dashboard. The part should now look like Figure 41 when in default orientation. The cut will be highlighted in red as usual, as the last feature created.

Saving the Part

It is a good idea when you are just getting started to save your model quite frequently, just in case something serious goes wrong. If you have to bail out of the program, you can always reload the most recently saved copy of the part and continue from there.

There are (as usual!) several ways to save the part:

- in the top toolbar, select the *Save* button, or
- in the pull-down menus select *File > Save*, or
- use the keyboard shortcut CTRL-S.

In the command/message window, you will be asked for the name of the object to be saved (remember that you can have more than one object loaded into memory at a time). Accept the default **[block.prt]** (this is the *active* part) by pressing the enter key or the middle mouse button. Pro/E will automatically put the part extension (*prt*) on the file. If you save the part a number of times, Pro/E will automatically number each saved version (like block.prt.1, block.prt.2, block.prt.3, and so on). Be aware of how much space you have available. It may be necessary to delete some of the previously saved versions; or you can copy them to a diskette. You can do both of these tasks from within Pro/E - we'll talk about that later.

> **IMPORTANT NOTE:**
> The *Save* command is also available when you are in Sketcher. Executing this command at that time will *not* save the part, but it will save the current sketch with the file extension *sec*. This may be useful if the sketch is complicated and may be used again on a different part. Rather than recreate the sketch, it can be read in from the saved file. In these lessons, none of the sketches are complicated enough to warrant saving them to disk.

Using Part Templates

You will recall that in the block part created earlier, the first thing we did was to create default datum planes. These (plus the named views based on them, which we didn't create this lesson) are very standard features and aspects of part files, and it would be handy if this was done automatically. This is exactly the purpose of part templates.

A template is a previously created part file that contains the common features and aspects of almost all part files you will ever make. These include, among other things, default datum planes and named views. Pro/E actually has several templates available for parts, drawings, and assemblies. There are variations of the templates for each type of object. One important variation consists of the unit system used for the part (inches or millimeters). Templates also contain some common model parameters and layer definitions[5].

A template is selected when a new model is first created. Let's see how that works. Create a new part (note that you don't have to remove the block - Pro/E can have several parts "in session" at the same time) by selecting

File > New

or using the "Create New Object" button. The **New** dialog window opens. The

Part | Solid

are selected by default. Enter a new name, like *exercise_1*. Remove the check mark beside *Use default template* and then select *OK*.

In the **New File Options** dialog window, the default template is shown at the top. It is likely "inlbs_part_solid". This template is for solid parts with the units set to inch-pound-second. It seems strange to have force and time units in a CAD geometry program. Actually, this is included so that the part units are known by downstream applications like Pro/MECHANICA which perform finite element analysis (FEA) or mechanism dynamics calculations. These programs are very picky about units!

Note that there are templates available for sheet-metal parts and for metric units (millimeter-Newton-second). While we are mentioning units, be aware that if you make a wrong choice of units here, it is still possible to change the units of a part after it has been created.

There are only two model parameters in the default template. *DESCRIPTION* is for an extended title for the part, like "UPPER PUMP HOUSING". This title can (eventually) be called up and placed automatically on a drawing of the part using, you guessed it, a drawing template. Similarly, the *MODELED_BY* parameter is available for you to record your name or initials as the originator of the part. Fill in these parameter fields and select *OK*.

The new part is created which automatically displays the default datums. They are even named for you (we will see how to name features in lesson 3): instead of DTM1, we have **RIGHT**. **TOP** replaces DTM2, and **FRONT** replaces DTM3. The part also contains a coordinate system, named views (look in the Saved Views List), and other data that we'll discover as we go through the lessons. The named views correspond to the standard engineering views. Thus, it is important to note that if you are planning on using a drawing template (discussed in a later lesson), your model orientation relative to the default datums is critical. The top-front-right views of the part are the ones that will be automatically placed on the drawing later. If your model is upside down or backwards in these named views, then your drawing will be too. This is embarrassing and not likely to win favor with your boss or instructor!

---

[5] Model parameters and layers are discussed in the *Advanced Tutorial*.

Now, having created this new part, you are all set up to do some of the exercises at the end of the lesson. Do as many of these as you can. Perhaps do some of them in different ways by experimenting with your sketch orientation, Sketcher commands, and so on.

This completes Lesson #2. You are strongly encouraged to experiment with any of the commands that have been presented in this lesson. Create new parts for your experiments since we will need the block part in its present form for the next lesson.

In the next lesson we will add some more features to the block, discover the magic of relations, and spend some time learning about the utility functions available to give you information about the model.

Questions for Review

Here are some questions you should be able to answer at this time:

1. What is meant by a blind protrusion?
2. What is the purpose of the sketching reference?
3. How do you specify the name of a part?
4. Give as many of the Sketcher implicit rules as you can.
5. How do you save a part?
6. What is a template?
7. What is your system's default template?
8. Where does your system store your part files when they are saved?
9. What is meant by the *active* part?
10. How does Sketcher determine the radius of a fillet created on two lines?
11. What happens if you delete any of the constraints (H, V, etc.) on a sketch?
12. What happens if you set the thickness of a thin solid greater than the radius of a filleted corner of the sketch?

Exercises

Here are some simple shapes that you can make with a single solid protrusion. They should give
you some practice using the Sketcher drawing tools and internal rules. Choose your own
dimensions and pay attention to alignments and internal constraints. The objects should appear
in roughly the same orientation in default view. Have a contest with a buddy to see who can
create each object with the fewest number of dimensions. This is not necessarily a goal of good
modeling, but is a good exercise! Feel free to add additional features to these objects.

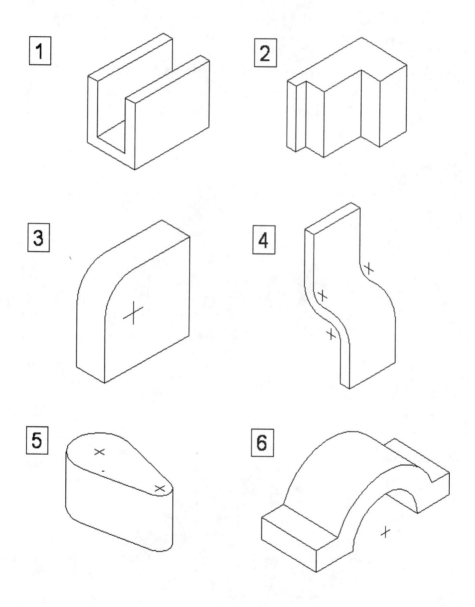

These parts are a bit more complicated, requiring two or more simple extruded features (protrusions or cuts). Think about these carefully before you try to make them.

1.

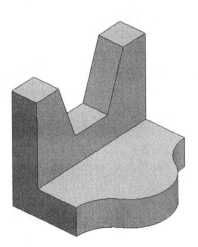

2.

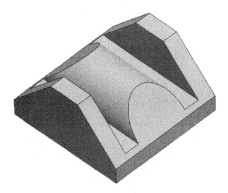

3.

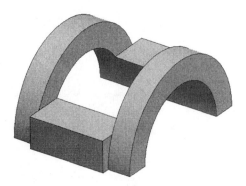

Lesson 3

Creating a Simple Object (Part II)

Synopsis

Placed features (hole, round, chamfer) are added to the block created in Lesson #2; customizing the model tree; naming features; modifying dimensions; adding relations to control part geometry; implementing *design intent* using relations; more Sketcher tools and options.

Overview of this Lesson

We will continue with the creation of the model you started in Lesson 2. We are going to add three new features to the block: a *hole*, a *chamfer*, and a *round*. These features do not require Sketcher since their geometry is more-or-less predetermined. We only have to specify where they go - they are *placed* on the model, rather than sketched. Then, we will explore some more of the interface tools using the model tree. We will modify some part dimensions, and then introduce the use of *relations* to adjust the geometry automatically. Finally, we will open up the sketch of the cut we made last lesson and look at how a feature's dimensioning scheme is used to implement *design intent*. Along the way we will come across some new tools and functions in Sketcher.

When we are finished this lesson, the block part should look like Figure 1. Although not obvious from the figure, there are a number of different ways we can create this simple geometry. This goes to the subject of *design intent*, which will be discussed towards the end of the lesson.

Here are the major steps we will follow, which should be completed in order:

1.   Retrieving a Part
2.   Adding a Hole
3.   Adding a Chamfer
4.   Adding a Round
5.   Customizing the Model Tree
6.   Naming features

Figure 1 Final part geometry

7.   Modifying Dimensions
8.   Adding Feature Relations
9.   Implementing Design Intent in Sketcher

We will be seeing a lot of new menus and dialog windows here. As usual, we will not discuss all commands and options in detail although some important modeling and Pro/E concepts will be elaborated. You should quickly scan each new menu/window to familiarize yourself with the location of the available commands and options.

Retrieving a Part

If you haven't already, login to the computer and bring up Pro/Engineer. If you are already in Pro/E, make sure there are no parts in the current session (select *File > Erase > Current*; then select *File > Erase > Not Displayed* to remove any other parts in the session).

Helpful Hint
If you need to change the default directory, use the commands:
File > Set Working Directory
and select the path to the desired directory for your *block.prt* file from the last lesson.

You can retrieve the block part we worked on last lesson using one of the following command sequences:

➤ in the pull-down menus, select *File > Open*

OR ➤ use the "Open" shortcut button on the top toolbar

OR ➤ use the Folder Browser in the Navigator.

Files in the current working directory are listed and there is a ***Preview*** button or function. This will be useful when your directory starts to fill up with part files by making it easier to select the file you want (especially if you are not very careful in using descriptive file names!). Note that the dynamic view controls (spin, zoom, pan) work in the preview window.

Note that it is easy to customize the displayed list of files by requesting only part files, assembly files, drawings, and so on. You can also change a setting that allows you to see all versions of a file (see the section *Pro/E Files Saved Automatically* at the end of this lesson).

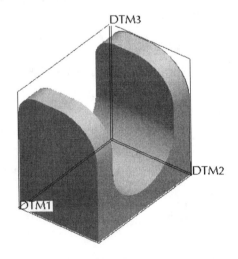

Figure 2 The **block** part at end of previous lesson

With **block.prt** highlighted in the file list select the ***Open*** button (or middle click). Pro/E will bring the part into the session, as shown in Figure 2. Close the Model Tree window if it is currently open and position the part in the default orientation.

Creating a Hole

The next feature we'll add to the block is the central hole. The major difference from the extruded features (protrusion and cut) we made before is that those were *sketched* features whereas a hole is a *placed* feature. It's shape is pretty much already defined, and all we have to do is tell Pro/E its size and where to place it on the model. Some other examples of placed features are rounds, chamfers, shells, pipes, and draft features. We will add some of those later this lesson. Although it sounds like a hole feature should be simpler than an extrusion, there are many variations of this feature. Some examples are shown in Figure 3 below.

A *straight* hole is a simple cylindrical hole with a flat bottom, essentially what you get with an end mill. A *sketched* hole involves the use of Sketcher to define the hole cross sectional shape or profile. This shape is revolved through 360° to create the hole. This obviously gives considerable freedom in the hole geometry which is handy for holes with several steps or unusual curved profiles. The *standard* holes can be countersunk, counterbored, neither, or both! Notice the shape at the bottom of the holes. Standard hole sizes are built-in for common bolts and thread specifications, and can be either tapped or clearance holes. If you pick a common thread specification, this will automatically create a note (that can be included in a drawing, for example)[1].

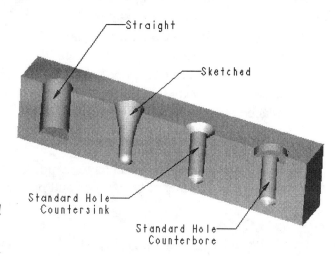

Figure 3 Example hole types

After its type and diameter, the next important variation in hole geometry is its depth. This is defined using one of the depth specifications shown in Figure 4. These are essentially the same as we saw previously for the cut feature. The *Blind* option drills the

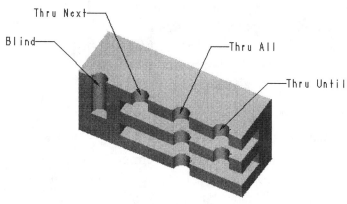

Figure 4 Hole depth options

---

[1] The helical thread is not actually shown on the hole, only a schematic representation. Showing the actual threads on dozens of holes would slow down the graphics display hardware!

hole to a specified depth. *Thru Next* will create a hole until it passes through the next surface it encounters. A *Through All* hole, as you would expect, drills through everything. Finally, a *Thru Until* hole goes up to a designated surface, edge, or point. If the hole is created "both sides" from the placement plane, then the hole depth can be defined separately in each direction.

A final requirement for placing the hole is the method for specifying where the hole is to be created, that is, its placement. As shown in Figure 5, this can be a *linear* dimensioning scheme (on the left) or a *radial* dimensioning scheme (on the right). Linear placement will position the hole using linear dimensions from selected references to its center point. The references are typically surfaces of the part or datum planes. Radial placement requires an axis, a radial distance from the axis, and an angular distance from a planar reference. Another common placement option (not shown) is *coaxial*, where the center of the hole is placed on an existing axis line.

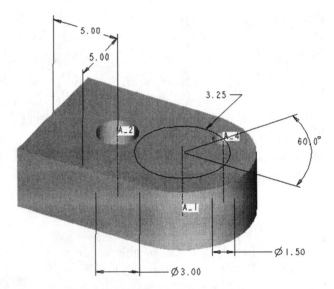

Figure 5 Hole dimensioning schemes: linear (left) and radial (right)

The **Hole** command is one of the more automated commands in Pro/E, and a lot happens here very quickly. Go through this slowly, and take time to explore some of the options and pull-up menus.

In the right toolbar, select the **Hole** command in the placed features toolbar. This will open the dashboard shown in Figure 6. The default is a simple, straight hole. The default depth is blind, that is, with a specified depth. The **Depth Spec** button on the dashboard lets you easily change this.

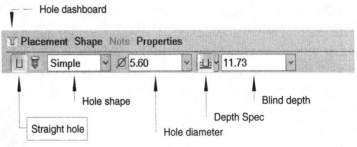

Figure 6 The *Straight Hole* dashboard. (default hole)

The alternative to a straight hole is a Standard hole, obtained by selecting the second icon from the left (see Figure 7). A standard hole lets you specify a standard screw size, thread type (UNC, UNF, etc.), a thread depth, tapping toggle. The nominal hole diameter is set automatically

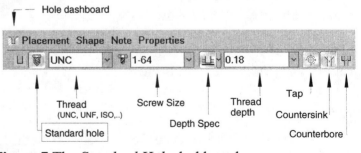

Figure 7 The *Standard Hole* dashboard.

when you pick the screw size. It also lets you add a counterbore or countersink to the hole. All the hole information is placed in a note which becomes part of the feature definition. When drawings are created of this part, the hole notes can be added. What is in the *Note* pull-up menu? Compare this to the *Properties* pull-up.

For a straight hole, another option for the hole shape is obtained using the *Sketched* profile option, Figure 8. Instead of a straight cylindrical hole, this lets you define the cross sectional profile of the hole (on one side since it is axisymmetric) using Sketcher. This is useful for creating stepped holes. An example of a sketched hole is discussed in Lesson #11.

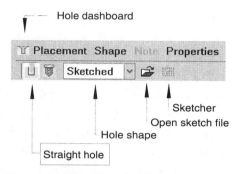

Figure 8 The *Sketched Hole* dashboard

For now, all we want is a *Straight, Through All* hole using *Linear* placement on the front surface of the block. This is the easiest hole imaginable to create that uses almost all default settings, so there is not much we need to change in the dashboard. As you perform the following, follow the prompts in the message window.

Click on the front face of the block at the approximate location of the hole center (mid-way between the left and right faces, 2/3 of the way up from the bottom). This surface is called the *primary reference* for the hole. You do not have to be very accurate with this since we will be setting exact dimensions next.

A hole preview will appear as shown in Figure 9. Notice that Pro/E has automatically figured out

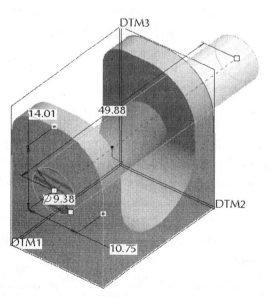

Figure 9 Default hole preview showing drag handles

which way the hole should go[2]. This feature preview has five drag handles whose functions are as follows (you may have to zoom in on the hole to distinguish between these):

- ▸   center location
- ▸   diameter
- ▸   distance to reference #1
- ▸   distance to reference #2
- ▸   depth

You can modify the hole geometry and placement by clicking on any of the drag handles. If you accidentally click on a surface instead of a drag handle, the hole primary reference (placement surface) will change. Not to worry - just left click back on the front face of the block. Try moving the center of the hole and changing its diameter or depth. The default placement type is *linear*. For the location drag handles, Pro/E wants you to select two edges, axes, planar surfaces or datum planes for linear dimensions to locate the hole center. We will use the right and top surfaces of the block for these references. All you have to do is drag the appropriate handle and drop it on either the right or top surface of the block (the surface will highlight in red when it is selected), as in Figure 9 above. Each time you select a surface, be sure to click on the *surface* not the *edge*. Once you have attached the handles to the references, double click on the dimension values to change the diameter to **10** and the linear distances to **10** from the right surface and **15** from the top (to center the hole on the front face). To change the depth from a blind hole to a through all hole, use the ***Depth Spec*** button on the dashboard. The hole preview will stop at the back surface of the part; the depth dimension disappears.

This completes the definition of the hole. Before you accept the feature, select the ***Verify*** button to see the hole. Also, examine the contents of the **Placement** and **Shape** pull-up menus to see how other hole options may be selected. As you move the cursor over reference fields in these menus, the appropriate edge/surface on the model will highlight.

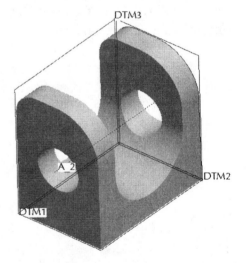

If all is well, accept the feature. If the preview shows something wrong, you can correct any of the element definitions by selecting the appropriate area or data field in the dashboard and making your corrections. Assuming the hole is correct, accept the feature. Your block should now look similar to Figure 10. Note that an axis line has been added automatically down the center of the hole (A_2 in Figure 10).

Figure 10 Hole created

---

[2]You might ask yourself how Pro/E knows this!

Creating a Chamfer

A chamfer is another example of a placed feature - the only
locational reference it needs is the edge (or edges) on which
it will be placed.

We will use the object/action command style here. Using
preselection (pick object - surface - edge with the left mouse
button), pick the two edges of the block shown in Figure 10.
Remember that to pick multiple entities, after the first is
selected you can hold down the CTRL key to include
additional entities in a selection set. It doesn't matter which
edge you pick first. When the edges are selected, they will
be highlighted in red.

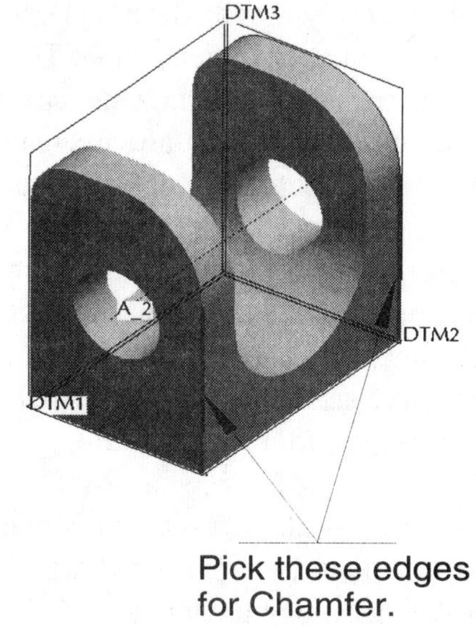

With the edges highlighted, pick the *Chamfer* tool in the
right toolbar. Be careful to get this one and not the *Round*
tool button which looks very similar.

A lot of things will happen on the screen at once. First, the
Chamfer dashboard will open at the bottom of the screen

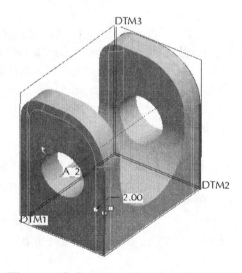

Pick these edges
for Chamfer.

Figure 10 Picking two edges
simultaneously for chamfer

(Figure 11). Also, the edges of the chamfer will be shown in preview yellow on the model. Two
drag handles will be attached to these edges. See Figure 12.

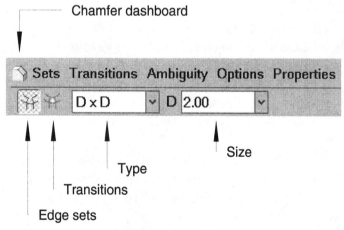

Figure 11 The *Chamfer* dashboard

Figure 12 Creating the chamfers

Let's explore the dashboard a bit. The chamfer icon is on the top left corner. In the bottom row,
the two icons at the left end determine the dashboard mode of operation. In "edge sets" mode,
you determine the collection or set of edges that will all have the same chamfer properties. It is
possible to create multiple chamfer sets within the same feature, each with different properties.
The next button is used to enter "transition" mode. As the icon indicates, this mode is used to

determine the geometry of chamfers that intersect or meet at corners. There are a number of ways that transition geometry can be set up[3]. None of these are necessary here since we do not have intersecting chamfers. The first pull-down list contains the options for setting the chamfer dimension type. The default (DxD) is an equal leg chamfer. The final text field contains the value of the indicated dimension D. This value is also indicated on the screen next to the previewed chamfer. You can change the chamfer dimension either by moving the drag handles, double clicking the value on the screen and typing in a new value, or by entering a new value in the dashboard. Set the chamfer distance to **2**.

Before you accept the chamfer, have a look in the pull-up menus on the dashboard. The **Sets** menu shows our current set, with dimension type and value indicated. If we had multiple sets, they would be listed here. The **Transitions** pull-up is currently empty.

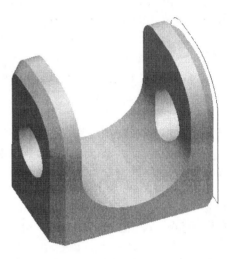

We have basically used all the defaults to create this equal length chamfer. Note that Pro/E has placed the chamfer by following along the tangent edge chain starting from the single straight edges we first selected. Use the *Verify* button on the dashboard at the right to see how the final geometry will appear. Finally, *Accept* the feature. The block should now look like Figure 13.

Figure 13 Chamfers complete

Creating a Round

Rounds are very common placed features that are created in the same way as a chamfer. Rounds are normally considered *cosmetic features*, and are therefore added to the model quite late in the regeneration sequence. Because they are so easy to create and make the model look more realistic (in shaded mode or rendered images), newcomers tend to go overboard in adding rounds to their models. However, be warned that misuse of rounds can have three detrimental effects on the model.

The first is their appearance in drawings. If a model has many rounds, the display of tangent edges showing the round extents can clutter up the drawing considerably. See the left side of Figure 14. On the other hand, if the display of tangent edges is turned off (which is standard practice), then important information about the part may sometimes be lost, as in the right side of Figure 14. The solution is to temporarily remove the rounds from the model for the drawing.

---

[3]Chamfer and round transitions are discussed at length in the *Advanced Tutorial*. We will not investigate them here.

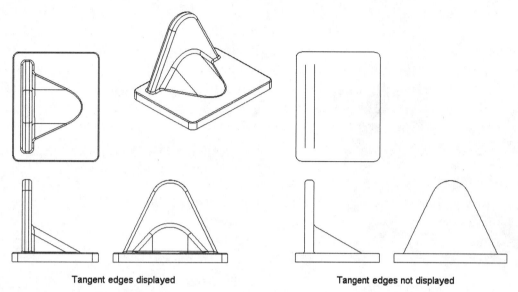

Tangent edges displayed   Tangent edges not displayed

Figure 14 Problems caused by rounds in drawings

The second effect of rounds is felt if the model is used for Finite Element Analysis (FEA). In FEA, the presence of rounds leads to a substantial increase in the modeling effort, primarily the number of elements required to represent geometry. This can drastically increase the model size and computational cost. Rounds do not usually have a large effect on the stress magnitudes in a part (unless they are at critical zones, especially fillets), so this increase in modeling effort is wasted (or may preclude the FEA modeling entirely). FEA models are often "de-featured" by eliminating unnecessary cosmetic features like rounds.

Finally, when a lot of rounds are present on a model, it is easy to accidentally pick a tangent edge as a reference for a subsequent feature. Creating this parent/child relation will usually cause problems.

In these cases (drawings, FEA, or avoiding inadvertent references), it will be handy to have the rounds organized in the model near the end of the regeneration sequence, and in some sort of orderly presentation so that they can be easily temporarily removed from the model (*suppressed*, see Lesson #5).

We have added all the key features to our block model, so now is a good time to add a final round feature. Using preselection, pick the edge shown in Figure 15. Now select **Round** toolbar button at the right (be careful not to pick **Chamfer**, which looks almost the same).

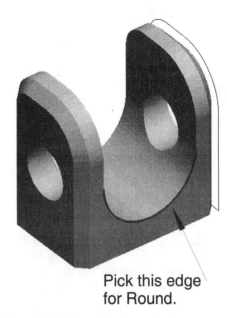

Pick this edge
for Round.

Figure 15 Creating a round

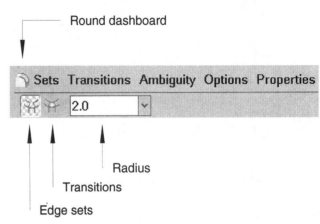

Round dashboard

Radius

Transitions

Edge sets

Figure 16 The ***Round*** dashboard

The **Round** dashboard appears as shown in Figure 16 (note the icon on the top row) and the round is previewed on the model (Figure 17) with yellow lines, a couple of drag handles, and a dimension (the round radius). The default round follows the tangent edge chain from the selected edge all the way around the model.

The dashboard looks and operates much the same as the chamfer dashboard. Selected edges to receive rounds are organized into edge sets (bottom row, left icon). A single round feature can contain several edge sets. You can use the transitions options to specify the type of geometry where rounded edges meet at corners.

Open the ***Sets*** pull-up menu to see the default settings for this edge/set. The round is *circular* (as opposed to *conic*, ie elliptical). The shape is a *rolling ball* round. Read the tool tip pop-ups for descriptions of these settings. The references pane lists the edges in the present set. The ***Transitions*** pull-up is currently empty. The ***Ambiguity*** pull-up (in much more complicated round cases) will list problem situations that you will have to resolve.

Even for simple rounds (with no transitions), you will find that there are considerably more options than there were for chamfers. A major variation available is the ability to create rounds whose radius changes along their length (*variable radius rounds*). The round extent can also be determined by other features (edges or datum curves). In this tutorial, we do not have time to explore all these options - you could spend an entire book chapter just studying rounds[4]!

---

[4] The *Advanced Tutorial* spends half a chapter on rounds, including variable radius rounds, surface rounds, and transitions, and still does not cover all the options.

Figure 17 Previewed edges and drag handles for simple round

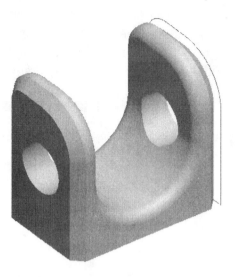

Figure 18 Completed round

For our purposes, this round is fine, so *Accept* the feature. The model should now look like Figure 18.

Exploring the Model

Configuring the Model Tree

Open the Navigator window. The model tree should be displayed. If not, pick the left tab on the Navigator menu.

In the model tree, click the "+" signs beside the protrusion and the cut. The tree now lists the sketches which were used with each of these sketched features. The other features do not have this because they were placed. The bottom entry in the tree is the *insertion point*. The next feature added to the model will be added to the tree here. We will find out in Lesson 5 how to move the insertion point around in the model, and what effect that has.

Review the features currently listed in the model tree. This is the regeneration sequence. Selecting a feature in the model tree will cause it to highlight on the model in the graphics window. To turn this highlight off, click anywhere on the graphics window. You may have to use the *Repaint* button on the top toolbar if your screen gets a bit cluttered.

To give us some more information about the features, we would like to add some columns to the model tree. Select the *Settings* tab just above the model tree and pick *Tree Columns*. This opens up the **Model Tree Columns** dialog window (Figure 19).

In the dialog window (Figure 19), have a look in the *Type* pull-down list. Select the default option *Info*. Then, in the left pane double click on **Feat #** and **Feat Type**. The *Width* control below the right pane will control the display of the selected column listed in the pane. Press the *Apply* button, then *OK.* The new columns are added to the model tree in the Navigator window (Figure 20). You might like to drag the right edge out a bit to see the entire tree. The column width can also be modified by dragging on the column separator bars

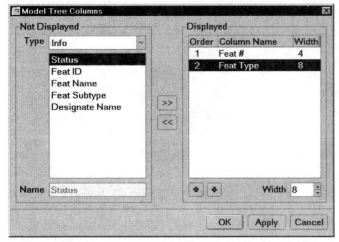

Figure 19 Configuring the Model Tree

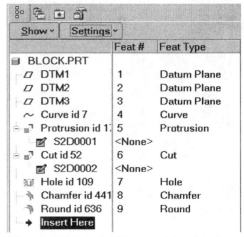

Figure 20 Model Tree with added columns

Select the *Settings* tab on the Navigator toolbar, then select *Tree Filters*. This opens the dialog window shown in Figure 21. Turn on the checks beside **Notes** and **Suppressed Objects**. It will be helpful while you are learning Pro/E to have everything turned on so that you won't miss anything. Open the other tabs (*Cabling, Piping, MFG*) to see what other information Pro/E can display in the model tree. Looks pretty complicated! Due to space and time constraints, we won't be dealing with any of these other items in this

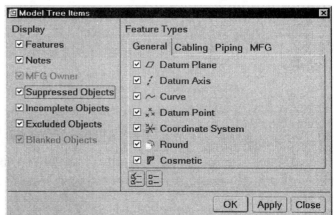

Figure 21 Displayed items in Model Tree

tutorial. Once again, *Apply* these settings (they won't change the model tree for this part) and select *OK*.

We want to save these settings so that the model tree will always be displayed this way. In the *Settings* tab, select *Save Settings File*. A dialog window will open showing your current working directory and any files with the cfg extension. These are configuration files. The default configuration file for the model tree is **tree.cfg**. This name is already in the file name text box at the bottom of the window. Go ahead and save this file now. The next time you launch Pro/E, the *tree.cfg* file will be read automatically and your model tree will display in the same way we have it now.

Naming Features

As was mentioned back in Lesson #1, it is a good idea to name the features in a model, or at least the major ones. This is important to help the model creator deal with large complex model trees. It is also indispensable for someone who must use a model created by someone else.

	Feat #	Feat Type
◻ BLOCK.PRT		
▱ RIGHT	1	Datum Plane
▱ TOP	2	Datum Plane
▱ FRONT	3	Datum Plane
∼ Curve id 7	4	Curve
⊞ BASE	5	Protrusion
⊞ U_CUT	6	Cut
⋎ BIG_HOLE	7	Hole
⌇ END_CORNER	8	Chamfer
⌇ INSIDE_CORNER	9	Round
➜ Insert Here		

Figure 22 Named features

Naming the features is quite easy. Select the feature in the model tree - it highlights on the screen. If you double-click on the feature in the model tree, the text field opens to let you change the default name to something more meaningful. Do that now with the features in the model tree, using names similar to those in Figure 22. The name is not case sensitive (as you type it in), and will be converted to all upper case.

Notice that even with only the name displayed in the model tree, the feature icon tells you what type of feature it is.

Helpful Hint

Just a reminder to periodically save your model. Now is a good time. Try using the keyboard command CTRL-S (hold down the Ctrl key and press S).

Exploring Parent/Child Relations

As you recall from Lesson #1, when a new feature is created, any previously created feature that it uses for reference is called a *parent* feature. The new feature is called a *child*. It is crucial to plan for and keep track of these parent/child relations. Any modification to a parent feature can potentially change one or more of its children. This may be precisely the desired model behavior and is the reason why we set up parent/child relations on purpose. However, with poor (or no) planning, changes to the parent may result in unintended (and unwanted) changes in the children. In the extreme case, deleting a parent will normally result in deletion of all child features that reference it (and their children...). In these cases, Pro/E will ask you to confirm the deletion. If you don't want to delete the child, you will have to change its references using techniques discussed in Lesson #5. Sometimes, if you are not careful about modifications to a parent, the child will be unable to regenerate. This is a symptom of a poor feature selection and/or referencing scheme. So it is important to be aware of what parent/child relations are present when new features are added. Be aware of the intent of your part geometry and build the model accordingly.

As you might expect, parent/child relations can become quite complicated when the model starts to accumulate features. This is a good reason to keep your models as simple as possible, and to think about your modeling strategy *before* you start creating anything! A parent can have many children, and a child can have several parents. Choosing (dare one say *designing?*) the best parent/child scheme for a part is a major difference of Pro/E from previous CAD programs. A clearly thought-out and implemented parent/child scheme is key to having flexible and robust models. Poor planning of the model will almost guarantee big problems later on if the model must be changed in any way. Fortunately, Pro/E provides a number of utility functions to help you manage the parent/child relations in a model. These include changing the dimensioning scheme and/or replacing current relations with new ones (called *rerouting*). In the worst case, reference elements of a feature can be *redefined*. We will be discussing these functions at length in Lesson #5. For now, you might keep as a general rule that, as in many things, simpler is better.

Let's explore the parent/child relations of the cut. Select this in the model tree (or use preselection to select it in the graphics window). Now hold down the right mouse button and in the pop-up menu select

Info > Parent/Child

This brings up the **Reference Information Window** shown in Figure 23. This shows the current feature at the top, the parents pane at the left, and the children pane at the right. Click on the parent BASE listed in the left window. The three surfaces of the feature will highlight in red on the model (see Figure 24).

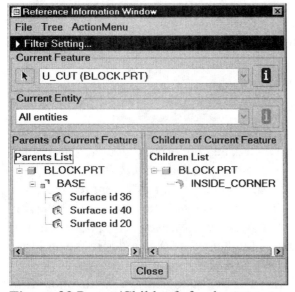

Figure 23 Parent/Child refs for the cut

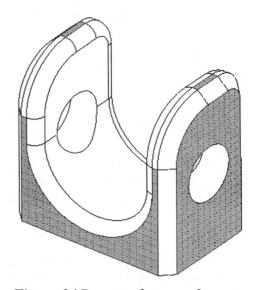

Figure 24 Parent references for cut

Expand the list by clicking on the "+" sign. Each of the three surfaces is listed. If you select these in turn, they will highlight on the model, and a short description of the nature of the reference will show up in the message window at the bottom of the screen. Look for the sketching plane and the section dimensioning references.

There is only one child of the cut - the round on the edge.

The feature *BASE* must have other children as well. Select it in the parent pane, hold down the right mouse button and select ***Set Current***. The feature *BASE* appears in the Current Feature area at the top. It has several children but it is interesting to note that the round does not appear on the child list. We already know that the round is a child of the cut. Evidently, the parent/child relations explicitly reported by Pro/E are only one generation deep (no grandchildren or grandparents!).

The *BASE* feature has three parents (the default datum planes), and several children. Note that the curve (feature #4 in the model tree) is not listed as a parent of the base feature. You recall we used this curve to specify the shape to be extruded to make that first protrusion. Also recall that we were warned that although the curve was used to create the shape, the curve feature would not be associated with the protrusion. This is now clear - the curve is *not* a parent. The shape of the curve has been copied into the sketch attached (in the model tree) to *BASE*.

Select the arrow button beside the **Current Feature** display field, and pick on the blue curve at the back of the block. This is feature #4 that was used to specify the protrusion. Note that it has three parents but no children listed. We should be able to delete it without affecting the model. We will do that in a minute or two after a couple more explorations.

Close the **Reference Information Window**. You should become comfortable using this window since it will be important in deciphering model structure.

Put the model back in default orientation.

Modifying Dimensions

There are numerous ways that you can change the shape of the model. You will do this most often by modifying its dimensions, as we saw briefly in Lesson #1. There are (at least) three ways of doing this. Let's experiment with the sketched datum curve. (We can mess around with this all we want without worrying about damaging the model, since the curve has no children!)

In the graphics window, double-click on the blue sketched datum curve. All the dimensions used in the sketch are shown. Put the mouse cursor over the line on the right edge. The cursor takes the shape of an open hand. Holding down the left button causes the hand to close and "grab" the line. Now dragging the mouse will cause the line to move. The movement of the line must be consistent with the existing dimensions and constraints on the sketch - tangency points must be maintained, vertical lines stay vertical, and so on. The only thing that changes is the dimension for the width of the sketch. Drop the line by releasing the left mouse button. What happens if you grab

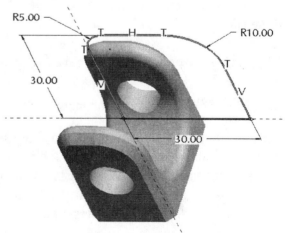

Figure 25 Modifying dimensions of curve

one of the corner fillets? When you drag an edge and release the mouse, what color is the affected dimension? If you click directly on the dimension value, you can type in precisely the desired value. Do that now, using the values shown in Figure 25.

If you left click somewhere else on the screen, the yellow dimensions disappear and the blue curve is shown in its original shape. Double click on the curve again. The shape with modified dimensions is still there.

New dimensions do not take effect until you *Regenerate* the model, using the toolbar button at the right end of the top toolbar. The new curve displays in blue, without any change in the protrusion (again indicating that it is not a parent of the protrusion).

Two other ways of modifying dimensions involve selecting the feature (do that now with the curve) in either the model tree or on the screen and holding down the right mouse button. In the pop-up menu that appears, select *Edit*. This will display all the dimensions as before. Double click on a dimension to change its value. Don't forget you have to *Regenerate* the model for the new value to take effect. What happens if you enter a negative number for the height or width dimensions? What is the value the next time you launch the *Edit* command?

We have found that we don't need the sketched curve in the model. Select it (either in the model tree or on the screen), hold down the right mouse button, and select *Delete* from the pop-up menu. You will have to *Confirm* this deletion.

We would like to make the *BASE* feature wider. Select it (double click on the end surface in the graphics window) and change the dimension 20 to a new value of **30**. *Regenerate* the part. It should look like Figure 26. Notice that the other features have all adapted to this change in the base feature - the cut is longer (it was a *Through All* cut), and the chamfers and round have also lengthened. The hole has maintained its distance from the right reference surface of the block, just the way it was dimensioned.

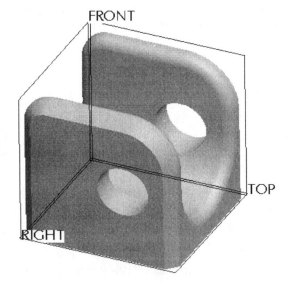

Figure 26 Modified *BASE* feature

What can go wrong here? Before we experiment, you should probably save the part. Now, double click on the *BASE* protrusion again and notice the size of the two rounded corners on top of its sketch. What happens if we try to make the sketch narrower than the sum of these two radii? We should expect trouble if we do that. Try doing it - say by entering a value of 12.5 for the width of the sketch. Zoom in on the top of the sketch for the protrusion and observe the shape. What happens when you *Regenerate*? You will be able to back out of this error by selecting *Undo Changes* in the **Resolve Feat** menu at the right.

You might try some "silly" dimensions (for example, make the diameter of the hole bigger than the height of the block), to see what Pro/E will do - in particular, what messages does it give you? Try changing the location of the hole so that it is completely off the left end of the block. As long as datum axes are turned on, you can still find it using preselection. What happens if you decrease the height of the *BASE* feature to 20? In particular, what happens to the round? If you get into serious trouble here, just erase the part from the current session (*File > Erase > Current*), and retrieve the previously-saved part (You did save it, right?) from disk (*File > Open > block.prt*). Before you proceed, return the dimensions to their original values and *Regenerate*.

Can we set up the model to ensure that the hole always stays in the center, regardless of the width of the block? We'll find out in the next section.

Note that when the dimensions are changed, Pro/E will still maintain all the geometric constraints that you set up during feature creation. A simple example of this is alignment or snapping to references - the edges of the cut were aligned with the upper surface of the block, for example. If the block height is increased, the top edges of the cut sketch move with the top surface of the block. If a feature is *completely* defined by this type of constraint (ie. all geometry is defined with alignments to references), then you will not be able to modify it directly by its dimensions since it has none! We saw this in a cut feature in the *tut_base* part in Lesson #1. You will only be able to affect it via its parent(s) dimensions.

The type of constraints on the geometry discussed in the last paragraph are built into the model during sketching of the feature. There is another way that we can define dependancy between dimensions of different features that is equally powerful - these are feature relations.

Creating Feature Relations

A *Relation* is an explicit algebraic formula that allows a dimension to be automatically computed from other dimensions in the part (or in other parts in an assembly) or from a numeric formula. This is an important way of implementing design intent. We will set up two simple relations to ensure that the hole in the block is always centered on its width, and mid-way between the top and bottom faces.

Turn the datum planes and axes off, and put the model in the default orientation. If you double-click on the hole, you will see its location dimensions: 10 from the right surface, 15 from the top surface. What we are going to do is create two formulas that will be used to compute these values based on the current dimensions of the block. There are a couple of ways to do this. One way is very quick, but you won't see the available options or commands for dealing with relations. So, we will do it the long way around first! From the pull-down menus, select

    Tools > Relations

The **Relations** dialog window opens. The relations toolbar buttons are shown in Figure 27. The most important of these are circled: *Switch Dimensions*, *Function*, *Insert Dimension Symbol from Screen* (this essentially means *Pick from Screen*).

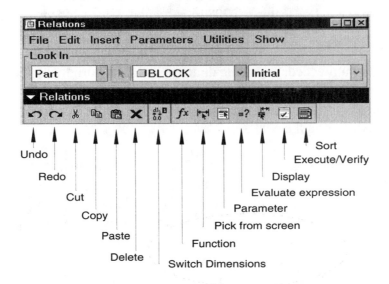

Figure 27 The *Relations* dialog window

Also, the dimension text on the hole showing the numerical values has been replaced with some symbols "dxx". See Figure 28. The "d" indicates a dimension; the number "xx" is a unique number defined by Pro/E, basically in the order that dimensions are added to the part.

The *Switch Dimensions* button in the **Relation** window toolbar is a toggle that switches between symbolic and numeric display of dimensions. Try that now. Note what the dimension symbolic names are in your model - the dimensions **d16** and **d17** in Figure 28 - for locating the hole. Your symbolic names may be different.

We need to find out the symbolic dimensions for the width and height of the protrusion. In the **Relations** toolbar, select *Insert Dimension Symbol from Screen*. Then (using preselection if necessary), click on the front surface of the block. Its dimensions should now also show up on the screen. Take note of the symbolic names for the width and height of the block (dimensions **d6** and **d7** in Figure 28). Middle click to close the **Select** window.

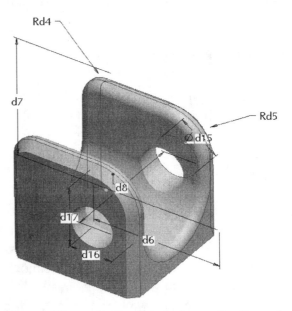

Figure 28 Symbolic dimensions of hole and protrusion

IMPORTANT: Your dimension labels might be numbered differently from these. Make a note of *your* labels!

Helpful Hint
Make sure the *Look In* setting at the top of the **Relations** window is set to *Part*. Otherwise, relations get attached to features and may be harder to find in the model later.

Now for the actual relations. In the text area in the **Relations** window, type in the following lines (see Figure 29) (**use your own dimension labels!**):

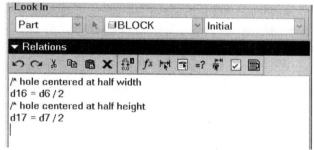

> */* hole centered at half width*
> *d16 = d6 / 2*

The first line of the relation, starting with **/***, is a comment line that describes the nature of the **Figure 29** Relations for centering the hole relation. This comment is not mandatory, but is a very good idea for clarity. You can put any text here that you like (multiple lines starting with /* are legal, too; blank lines are ignored). The next line defines the relation itself - the distance from the end face to the center of the hole is half the width of the block.

Here is another way to enter dimension symbols into the **Relations** window. Select *Switch Dims* so that they are showing up in numerical form. Now type in the next line of text:

> */* hole centered at half height*

In the next line, instead of entering the dimension symbol text directly, select the *Pick from Screen* button in the **Relations** toolbar. Pick on the hole dimension from the top of the block (15). The appropriate symbol (like **d17** in Figure 28) will appear in the Relations editor. Type in an equals sign, select the *Pick from Screen* button again, and pick on the height dimension for the block (30). The symbol (like **d7** in Figure 28) will appear. Complete the relation as shown in Figure 29. Notice that by using this tool, we don't need to first look up the symbolic names.

Select the *Execute/Verify* button in the relations toolbar. This will confirm that the relations are correct (or at least that they contain no syntax errors).

Before you leave the **Relations** dialog window, in the pull-down menu select

> *Show > Info..*

This opens a Browser page that lists all the relations in the part. It also reports the values computed by the relations. A lot of other information is presented that deals with mass properties of the part, which we will discuss later in these lessons. Close the Browser page, then accept the **Relation** dialog window with *OK*.

Note that the relations haven't taken effect yet. Select the ***Regenerate*** button on the top toolbar. The hole should move to the center of the block. Double-click on the block and change the width and height dimensions to *60* and *40*, respectively. ***Regenerate*** the part. If all goes well, the hole should be exactly centered on the block. Change the dimensions back to their original values (30 and 30). HINT: Use the pull-down lists of recently used dimensions beside each dimension field. ***Regenerate*** the part.

Double-click on the hole and try to change either of the dimensions that locate the hole - Pro/E won't let you! And it even tells you what relation is driving that dimension.

Figure 30 Centered hole using relations

More about relations:

Relations can take the following forms:

> **/* explicitly define a dimension**
> **d4 = 4**
>
> **/* explicitly define a parameter**
> **length_of_block = 30**
>
> **/* use a parameter**
> **d6 = length_of_block**
> **d12 = length_of_block / 2**
>
> **/* set up a limiting value for a dimension**
> **d4 > 2**

Explicitly defined dimensions are just that - they create constant values for dimensions that cannot be overridden. The right hand side of a relation can contain almost any form of arithmetic expression (including functions like sin, cos, tan, ...). The inequality form can be used to monitor the geometry during regeneration of the part. If the inequality is violated, then Pro/E will catch the violation and show you a warning message.

These are the simplest form of relations - simple assignments. There are a number of built-in functions that can be used in relations (like locating entities in a model, logical branching, and so on). Relations can be used almost like a programming language. For example, relations can be used to solve systems of simultaneous equations involving part parameters, or be used to perform design calculations that yield dimensional values. Very high level functionality can be obtained using a module of Pro/E called Pro/PROGRAM (see the *Advanced Tutorial*).

All the relations for a part go into a special database that is consulted/executed when the part is regenerated. These relations are evaluated in a top-down manner, so that the order of relations is important (just like the order of feature creation). You can't have two relations that define the same dimension, and a relation is evaluated based on the current values on its right hand side. If one of the right-hand side values is changed by a subsequent relation, then the dimension using the previous value will be incorrect. Pro/E has a utility function that will let you reorder the relations to avoid this. When re-ordering, Pro/E assumes that each relation is preceded by a single comment line that will be moved with the relation when the database is reordered.

Considering Design Intent

The notion of *design intent* arises from the fact that there are always alternate ways of creating the model. Even for our simple cut, there are a number of possible dimensioning schemes that would all describe the same geometry. We must choose from these alternatives based on how we want the feature to relate to the rest of the part (or to itself). This is called *design intent*.

Our design intent can be implemented by a combination of feature selection and creation order, explicit constraints, parent/child relations, feature relations, and the dimensioning scheme. We have seen some of these methods. Let's go back and see how the dimensioning scheme of a feature can be used to implement design intent. Use preselection to select the U-shaped cut. When this is selected, use the right mouse button to select *Edit Definition*. This opens the extrusion dashboard. On the far left, pick the *Sketcher* button and in the **Section** dialog that opens, select *Sketch*. This takes us to Sketcher with the sketch for the cut displayed (Figure 31). You might like to go to wireframe mode here and you can close the model tree.

Design Intent Alternative #1

The dimensioning scheme we used before expresses a design intent as follows:

- center of cut is 15 from the back of part
- radius of cut is 10
- center of U-shape is 15 from top of part

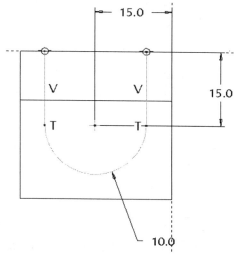

Figure 31 Sketch of cut with first design intent

Design Intent Alternative #2

Suppose we wanted a different design intent as follows:

- center of cut is 15 from the back of part (same as before)
- radius of cut is 10 (same as before)
- clearance from bottom of cut to the bottom of the part is 5

We saw previously that decreasing the height of the block caused the existing cut to pass through the bottom of the part. Our new intent is that the bottom of the U should always be exactly 5 from the bottom of the block. We can easily set this up using a different dimensioning scheme in the sketch.

In the toolbar at the right, select the ***Dimension*** button. Click once on the arc (not the center) and once on the bottom edge of the part. These will highlight in red. Move the cursor to the side and middle click at the location where you want the dimension text to be placed.

Since we are now over-dimensioning the sketch, the **Resolve Sketch** window will open (Figure32). Recall that Intent Manager will not ask for confirmation if it wants to delete a weak dimension - it just does it. All the existing dimensions, however, are strong. Intent Manager won't delete any strong dimension or constraint without asking. The **Resolve Sketch** window shows the conflicting dimensions and these are highlighted in red on the screen. The dimension currently selected in this window will have a yellow box around it on the screen. At least one of the three dimensions is no longer necessary. A dimension we can afford to lose is the 15 dimension. Select it and press the ***Delete*** button at the bottom[5]. The modified sketch with our new design intent implemented is shown in Figure 33. Accept the new sketch and the feature.

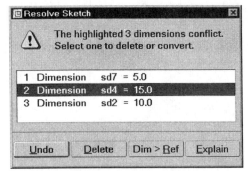

Figure 32 Resolve Sketch window showing conflicting dimensions

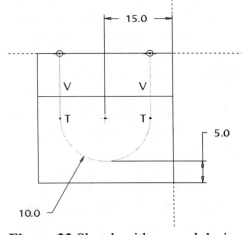

Figure 33 Sketch with second design intent

---

[5] You can also convert this to a *reference dimension*. Reference dimensions cannot be used to change the sketch geometry, but just indicate values in the sketch. In Pro/E jargon, they are *driven* dimensions rather than *driving* dimensions.

With this new design intent, change the height of the block to **20** and **_Regenerate_**. If you recall, this is the value we used before that resulted in the part splitting into two pieces. That won't happen this time, due to our intent (Figure 34). Change the height back to **30**.

Figure 34 Design Intent #2

Design Intent Alternative #3

Let's try one more variation on the design intent. Suppose we wanted the following:

- ensure a thickness of 5 between the vertical sides of the cut and the front and back surfaces of the block
- clearance from bottom of cut to the bottom of the part is 5 (same as previous)

Select the cut feature again, and use the right mouse button to select **_Edit Definition_**. Re-enter the sketch and pick the **_Dimension_** command again, and add the dimensions shown in Figure 35. You will have to deal with the **Resolve Sketch** window again, and delete some of our previous dimensions. You will also have the opportunity to delete constraints here (which we don't want to do).

Don't leave this sketch just yet - there are more tools to investigate.

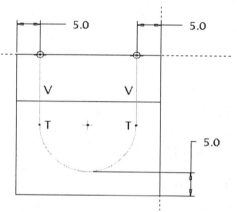

Figure 35 Sketch showing third design intent

More Sketcher Tools

The *Modify* Command

We have seen how to modify an individual dimension by double-clicking on it. You can also change the sketch by grabbing a sketched entity and dragging it with the mouse. Here is yet another way.

Click on the clearance dimension at the bottom of the cut. Now select the **_Modify_** button in the toolbar. The **Modify Dimensions** window appears (Figure 36). To the right of the dimension value is a thumbwheel. Drag this with the left mouse button. Experiment with the Sensitivity slider.

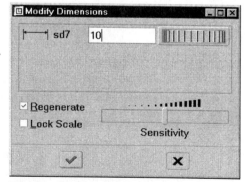

Figure 36 The **Modify** window

Return the dimension to its original value (**5.0**) by typing it into the data field and hitting *Enter*.

Now hold down the CTRL key while you left click on the other dimensions in the sketch. These will be added to the **Modify** window. Check the box beside *Lock Scale*. Now drag the thumbwheel beside any of the dimensions. All dimensions are changed simultaneously, in the same proportion.

Helpful Hint

The *Lock Scale* option is particularly useful when you are sketching the first feature in a part. Recall that the numerical values created for the first feature are chosen at random. If you try to modify the dimensions one at a time, you will probably destroy the shape of the sketch. If your sketch is more or less the right shape, you can change all the dimensions simultaneously using *Lock Scale* without changing the shape of the sketch. This is a great time-saver.

Remove the checks beside both *Lock Scale* and *Regenerate*. The latter option will delay the simultaneous and/or immediate regeneration whenever a single dimension is changed. This is sometimes necessary when you want to change several dimension values at the same time, but don't want to regenerate until all new values are entered. This would avoid trying to regenerate to a geometry with some old and some new dimensions, which might be incompatible. When you are finished experimenting, return the dimensions to the original values and close the **Modify** window (or select the X symbol).

Sketcher Relations

We would like to ensure that the thickness of the part at the front and back are equal. Using Design Intent #3 from above, we have two dimensions that we want to make equal. We could do that using a relation defined at the part level as we did before. There is another way here that is quicker.

While you are in sketcher, in the top pull-down menus select

Info > Switch Dims

The dimension labels on the screen will change to their symbolic values, with an "s" in front indicating they are sketch dimensions. Note the dimension at the top right (**sd8** in Figure 37).

To very quickly enter a relation, double-click on the thickness dimension on the top left (**sd9** in Figure 37).

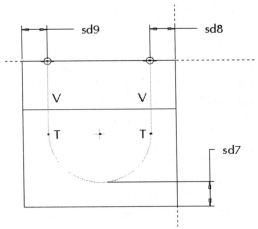

Figure 37 Sketch dimensions

Instead of typing in a numeric value, just type in the symbolic name of the other dimension on the top, **sd8** in Figure 37. The message window asks if you want to add the relation

$$sd9 = sd8$$

Middle click to accept this. Select

Info > Switch Dims

again to get the numerical display back. Now, try changing the value of the top left dimension. You can't. Change the value of the other dimension to **7.5**, as in Figure 38. Both dimensions change, indicating that the relation has executed.

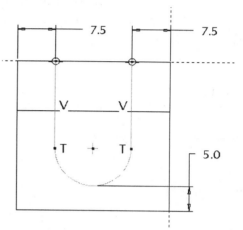

Figure 39 Final sketch for cut

Sketcher Preferences

There are a number of options for how you want Sketcher to behave. To investigate these settings, select (in the top pull-down menu)

Sketch > Options

This brings up a dialog window containing three tabs. Selecting the tabs will open the windows shown below. Come back to these and experiment with the settings later.

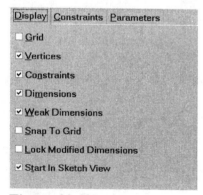

Figure 39 Sketcher options - *Display*

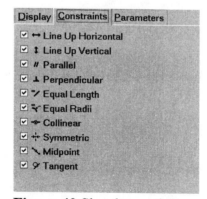

Figure 40 Sketcher options - *Constraints*

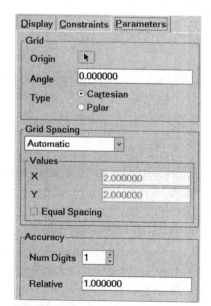

Figure 41 Sketcher options - *Parameters*

Using *Undo*

Another useful command in Sketcher is ***Undo*** ↶ located at the top of the screen. Each time you select this command, you move backwards through any changes you have made in the sketch, one at a time. You can move forward again (with some restrictions) using ***Redo*** ↷.

We have now finished with Sketcher for this lesson, so accept the sketch (it should look like Figure 38) and complete the cut feature.

Saving the Part

We are (at last!) at the end of this lesson - it's been a long one. Before you leave, make sure that you save the current part, that should look something like the figure at the right.

File > Save

or use Ctrl-S. You can now exit from Pro/E.

Figure 43 Final part geometry

Pro/E Files Saved Automatically

Have a look at the files in your default disk space or Pro/E working directory. You should see files listed that include the following forms:

block.prt.1   block.prt.2   block.prt.3   . . . .

Each time you save a part (or drawing or assembly), a new file is created with an automatically increasing counter. Thus, you always have a back-up available if something goes very wrong. On the other hand, this can eat up your disk space very quickly since the part files can get pretty large. If you are sure you do not need the previous files, you can remove them. Since it is always a good idea to keep back-ups, you might consider copying final part files to another storage location anyway (see your system documentation for this).

Pro/Engineer will also write a number of other files to your disk space. These might include the following:

trail.txt.??

This is a record of all keystrokes, commands, and mouse clicks you made during a session. For an advanced user, this may be useful to recover from catastrophic failures! Each new session you launch starts a new trail file, with an automatically incremented counter. These files are often stored somewhere else on your system, away from your working directory.

feature.lst
>The same list of features obtained using *Info > Feature List*

feature.inf
>The data about features obtained using *Info > Feature*

rels.inf
>All the dimension relations in the part or assembly file

reviewref.inf
>Information on parent/child relations

and other *.inf files.

Unless you have a good reason to keep these, remove them from your disk space as soon as you leave Pro/E (and not before!). There are a number of programs available for download from the Web (some free) that will automatically purge these files from your directories.

In the next lesson we will look at a number of new features, including revolved protrusions, mirrored copies, and more Sketcher tools that will extend our repertoire of part-creation techniques. In the meantime, here are some questions for you. Some review material we have covered and others will require you to do some exploring on your own.

Questions for Review

1. What elements are required to define a simple hole?
2. What is meant by linear placement of a hole?
3. What is the difference between the terms "protrusion" and "extrusion"?
4. What is a useful method for visualizing a new sketch relative to the part?
5. What are two methods of obtaining a feature list?
6. What commands are used to name features?
7. Suppose you have a very complex part with many features and you want to identify/locate (ie. show graphically) a specific feature in the model. How would you do it?
8. When might you want to turn off *Regenerate* in the **Modify Dimensions** dialog window?
9. Once a feature has been created, how can you change its dimensions?
10. What do we call an equation that computes a dimensional value?
11. What are the "junk" files that Pro/E creates in your disk space? How do you get rid of

them?

12. How can you go back and edit a previously defined relation?

13. How can you find out the internal symbolic names for feature dimensions?

14. What happens to the relations if you delete a feature whose dimensions appear (a) on the right side, or (b) on the left side of a relation?

15. What is the difference between the **Feature Number** and the **Internal ID**?

16. What is the first thing you have to do when you have entered Sketcher?

17. What is the meaning of gray dimensions? What about the yellow, red, green, and white ones?

18. How can you strengthen a dimension? What does this mean?

19. Where did we see the *Lock Scale* option? What does it do?

20. How do you over-ride the default dimensions placed by Intent Manager?

21. How do you change the location of the dimensions (on the screen) after they have been placed by Intent Manager?

22. What does the right mouse button do in Sketcher?

23. Find out how to turn off the constraints presented by Intent Manager (not just turn off the display, but actually get rid of them).

24. What is the minimum number of Sketcher References needed by Intent Manager? The maximum number? What do these do?

25. Are the following sets of references sufficient or not for defining the Sketch (can the sketch be "fully placed")?
    a.   a single vertical reference line
    b.   a single point at the center of a circle
    c.   a pair of parallel lines

26. In your own words, describe what is meant by "design intent." How was design intent implemented in the part created in this lesson.

27. Examine some simple everyday objects and describe how you might implement design intent in a computer model of the object. Make some freehand sketches to illustrate.

28. What happens if you turn the Intent Manager off part way through the creation of a sketch? How do you do that? Can you turn it back on again?

Exercises

Here are some simple shapes you should be able to make using the features covered so far. When complete, the shapes should be approximately in the positions shown in default view. Before starting in on any new part, take a few minutes to plan your modeling strategy. For example, where should the datum planes be located? This will pay dividends in the ease with which you can model the part, and particularly with how you will be able to modify it afterwards.

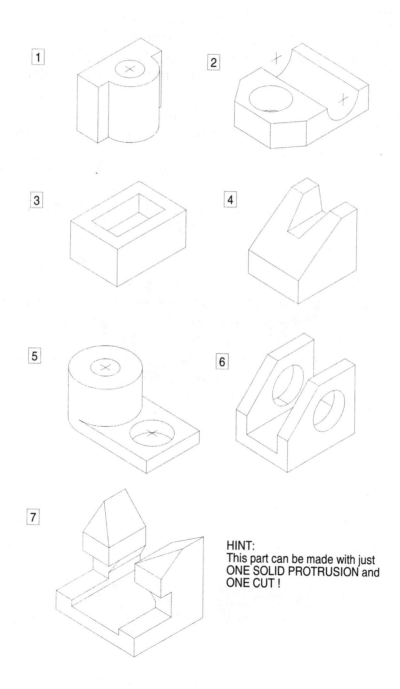

HINT:
This part can be made with just
ONE SOLID PROTRUSION and
ONE CUT !

These parts are a bit more complicated, and use more of the features covered this lesson (holes, chamfers, rounds).

8.

9.

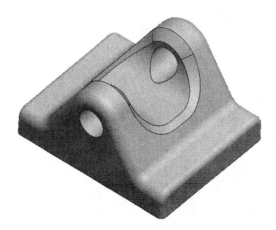

10.

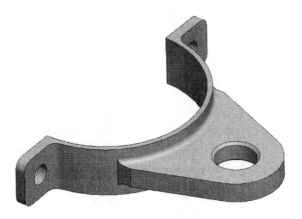

Project

Here is the first part in our assembly project. It uses only the features covered in the previous two lessons. All units are in millimeters. As usual, take a few minutes to plan your modeling strategy. For example, where should the datum planes be located? How should you orient the part in the Front-Top-Right system of datums (assuming you are using a template). Which is the base feature?

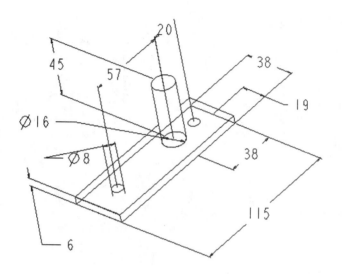

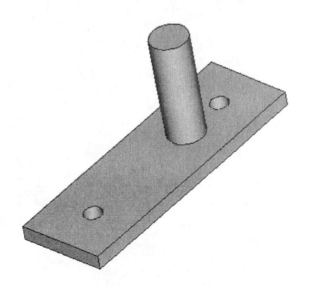

This page left blank.

Lesson 4

Revolved Protrusions, Mirror Copies, Model Analysis

Synopsis

A new part is modeled using a number of different feature creation commands and options: both sides protrusions, an axisymmetric (revolved) protrusion, a cut, quick rounds, and chamfer edge sets. More Sketcher tools. Mirrored features. Error recovery.

Overview of this Lesson

This lesson will introduce you to an important feature geometry (a revolved protrusion), and give you some practice using ones introduced in the first two lessons. Because Sketcher is such an important tool, we will spend some time exploring more tools and functions, and discussing how it can be used most effectively. The part modeling steps should be completed in order.

Remember to scan through each section before starting to enter the commands - it is important to know what the goal is when you are going through the feature creation steps. If you can't finish the part in one session, remember to save it so that you can retrieve it later and carry on. The finished part should look like Figure 1. Here are the steps:

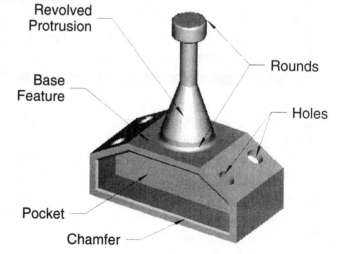

Figure 1 Finished Part

1. Creating the base feature
2. Adding a revolved protrusion
3. Adding a pocket with a cut
4. Adding holes
5. Adding rounds and chamfers
6. Model analysis tools
7. Exploring "What Can Go Wrong?"

IMPORTANT:

Be sure to complete the last section. You will learn a lot about how Pro/E works, beyond which button to push. This is important for your proficient use of the program.

As usual there are some Questions for Review at the end, some exercises, and another part for the project.

The instructions are going to be a bit more terse this lesson, especially for commands we have covered previously. You should be getting in the habit of scanning both the command menus and the message line in the command/message window. Remember, if the mouse seems to be dead, then Pro/E is probably waiting for you to respond to a prompt via keyboard entry. By now, you should also be fairly comfortable with the dynamic view controls obtained with the mouse.

So, get started by launching Pro/E as usual. Close the Browser and Navigator windows. Study the object in Figure 1 carefully before proceeding.

Create a solid part named **guide_pin** using the **Create New Object** button or select:

> **File > New > Part | Solid | [guide_pin]**

Use the default part template.

Creating the Base Feature

The rectangular block at the base of the part will be our first solid feature, also sometimes called the *base feature*. We will be creating the base feature so that the **FRONT** and **RIGHT** datum planes can be used for mirroring of features we will create later.

> **Helpful Hint**
> Whenever you have symmetry in a part, it is a good idea to use the datum planes on the plane(s) of symmetry. That way, they will be available for mirroring and serving as references for symmetric features.

Thus, we will create the first feature as a *blind, symmetric* extrusion: the sketch will be on FRONT and the extrusion will extend a specified distance on both sides of the sketching plane. Select the **Extrude** tool in the right toolbar. Activate **Sketcher** using the button on the dashboard (lower left corner). Now you need to select a sketch plane and reference plane. Choose **FRONT** as the sketch plane. The RIGHT datum is automatically chosen as the orientation reference. This is what we want, so just middle click to accept the dialog and enter Sketcher. The sketching references have been selected automatically for you, so you can Close that dialog window.

The base feature will be created symmetrically about **RIGHT**. This placement means that the axis of the vertical revolved protrusion we'll create later can be aligned with the vertical datum planes that cross in the center of the base. This is an example of the planning ahead you must do. This one was easy - only "one move ahead." Like good chess players, good modelers are always looking many "moves" ahead.

Before we start the sketch, recall the sequence we want to follow with Sketcher:

1.   make sure the desired references are selected;
2.   sketch the geometry using the chosen references for alignments, constraints, etc.;
3.   strengthen whatever dimensions or constraints on the sketch you want to keep;
4.   modify the constraints if required to implement your design intent;
5.   modify the dimension scheme so that it implements your design intent;
6.   modify the dimension values to those desired for the feature.

The sketch we want to create is shown in Figure 5. We'll get there in several steps, corresponding to the sequence just given.

The first step has been done for us - Sketcher picked the only two possible references for us. With these identified, you can turn off the datum planes and coordinate system, as they will not be needed for a while.

On to the second step. Use the right mouse button pop-up menu to select *Line*. Letting the cursor snap to the references, you can create the entire sketch using a single polyline (left click, left click, ... seven times). Middle click to leave *Line* mode. See Figure 2. Intent Manager will put all the weak constraints and dimensions on the sketch in gray. Since this is the first feature of the part, these will be chosen seemingly at random. Some will be the ones you want; others won't. DO NOT bother modifying these dimension values yet - this will result in wasted effort. What we are interested in first is getting the shape and proportions of the sketch right.

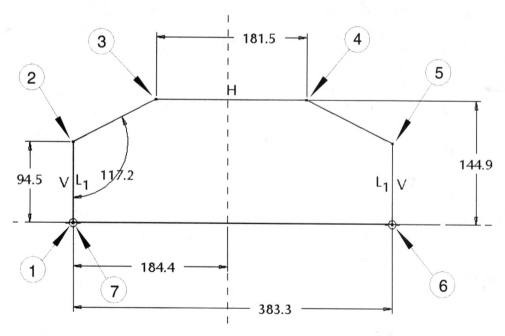

Figure 2 Creating the sketch for the base feature (constraints and dimensioning scheme not completed)

Now, move the dimensions off the part and observe the constraints and dimension scheme that has been created by Intent Manager. Compare these to the desired constraints shown in Figure 5. Our third step is to select aspects of the sketch we want to keep - constraints and dimensions. We do this by making them strong. Select one you want to keep. When it is highlighted in red, use the right mouse button pop-up menu and select **Strong**. It will change from the weak color (gray) to the strong color (yellow). Continue doing this for any constraints and dimensions corresponding to those in Figure 5. We are still not worried about dimension values.

Now we can implement any missing constraints that have not been deduced by Intent Manager.

We can implement a left-to-right symmetry about RIGHT (the vertical reference) as follows. Use the flyout on the **Line** button to select the **Centerline** button, *OR* right click on the graphics window to get the pop-up menu and select **Centerline**. Sketch a vertical centerline on the vertical reference. When the centerline appears (yellow dashed line), if your sketch is already close to being symmetric about this line, Sketcher may automatically apply the symmetry constraint. **Repaint** your screen to look for the small symmetry constraint arrows. These are weak (gray) so they may be hard to see. If these are missing, read on...

If your current sketch is missing some of the constraints shown in Figure 5, we can set some or all of these explicitly. Select the **Constraints** button in the right toolbar. This opens a window containing the nine explicit constraint options shown in Figure 3. Examine these carefully, and add any constraints on your sketch so that it matches the desired figure. For example, for the symmetry constraint, select the symmetry constraint button (lower left) and read the message window. Click on the vertical centerline and then the two lower vertices. Note that with symmetry constraining the sketch, the dimensioning scheme has probably changed. Check out the symbols that indicate the symmetry. If necessary,

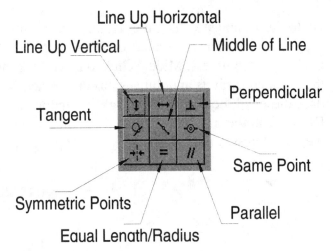

Figure 3 Explicit constraints

repeat this process for the two vertices on the top edge of the sketch. The sketch also contains an equal length constraint on the vertical edges. If required, add this to your sketch. The sketch constraints should now look like Figure 4.

So far, the shape and constraints are set the way we want, but the dimension scheme and values probably are not. Again, compare to the dimensioning scheme shown in Figure 5. If any of these dimensions are missing, create

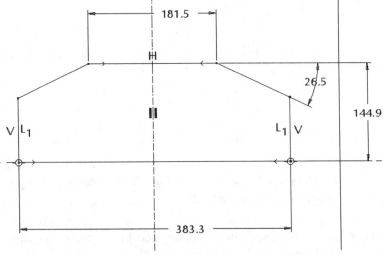

Figure 4 Sketch with desired constraints and dimension scheme

them explicitly. Pick the ***Dimension*** command from the right toolbar. Recall that these will be strong dimensions and Intent Manager will remove redundant weak dimensions automatically.

To dimension the angle, click on the two intersecting lines and middle click where you want to place the dimension text. Sketcher assumes that if two lines intersect you must want the angle between them.

Finally, once you have the shape, constraints, and dimensioning scheme you want, you can modify the dimension values. The initial dimension values chosen by Sketcher for the base feature are fairly arbitrary. If we tried to modify dimensions one-by-one, there are two possible problems. First, the shape of the sketch may become grossly distorted, to the point that Sketcher may not be able to solve it.. Second, it is possible that Sketcher might have trouble recomputing the sketch because we may be requesting incompatible values. What we want to do is keep the shape of the sketch the same, while scaling all dimensions the same amount. In the last lesson, we found a very useful command to do this. CTRL-click with the left mouse button to select all the linear dimensions (not the angle!). Select the ***Modify*** button. The three dimensions will appear in the **Modify Dimensions** window.

Now, check the ***Lock Scale*** option (since we want to change all dimensions simultaneously).

Select the dimension for the block width and enter **20** into the data field. The other dimensions will change at the same time in the same proportion so that the shape of the sketch is not damaged. Now uncheck the ***Lock Scale*** option. Enter new values for the other dimensions according to Figure 5. Finally, close this window and change the angle dimension (hint: double-click on the dimension).

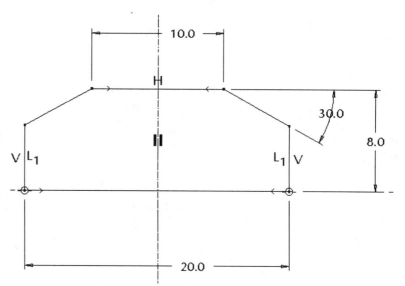

Figure 5 Final sketch for base feature

This should complete the sketch and it should look like Figure 5. So, select the ***Accept*** button at the bottom of the Sketcher toolbar.

You should review this Sketcher sequence again - using it properly can save a lot of frustration.

We now see a preview (in yellow) of the protrusion. The default is a one-sided, blind protrusion (turn the datum planes back on to see where our sketch was on FRONT). On the dashboard, open the **Depth Spec** pull-up list, and select the ***Both Sides*** option. The blind dimension value specifies the total depth (symmetric about the sketch plane). Enter a value of 10 (either in the dashboard or on the dimension shown in the graphics window). The preview should look like Figure 6.

You can now *Verify* the protrusion. Assuming everything is satisfactory, *Accept* the feature.

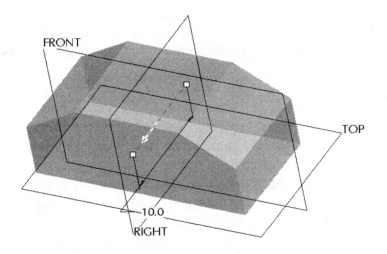

Figure 6 Base feature preview (*both sides, blind*)

Creating a Revolved Protrusion

We'll now add the vertical axisymmetric shape onto the top of the base feature. In 3D solid modeling terms, this is a "revolved solid", created by taking a 2D sketch and rotating it around a specified axis. In Pro/E, we can use revolved features to create protrusions or cuts. The angle of the rotation is adjustable. For this part we will do a 360° revolve. Only a half cross-sectional shape is required. See Figure 7. Depending on the model, the sketch can be either an open or closed curve. **The sketch must also include the axis of rotation**.

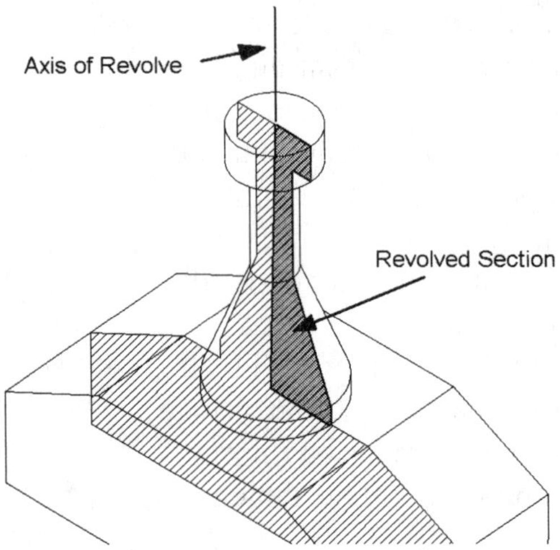

Select the *Revolve* tool in the right toolbar. The Revolve dashboard looks the same as the Extrude dashboard, and offers the same options (thin

Figure 7 Section to be revolved to form protrusion

feature, remove material, depth spec, and so on). Select and launch *Sketcher* from the dashboard. Select **FRONT** as the sketching plane. The default sketch orientation reference will be OK, so middle click or select Sketch.

The **References** window is now open, likely with RIGHT and TOP datum planes selected as references. The sketch we are going to create is shown in Figure 9. This is essentially the right half of the cross section of the revolved feature. We want the lower vertex on the sketch to lie precisely on the top of the block. The easy way to do this is to make the top surface a reference. You may want to spin the object to see this surface. (If you do, there is a button in the top toolbar to reorient your view of the sketch to the standard view.) Select the surface now (see Figure 8).

This will create another horizontal reference in the sketch. The first one (TOP) can be deleted. The reference status should still be "Fully Placed."

Create the sketch shown in Figure 9 (display of H and V constraints has been turned off for clarity). Remember the desired sequence for efficient use of Intent Manager:

- select the appropriate references (done that!);
- sketch the desired shape (doing that!)
- strengthen any constraints or dimensions you want to keep;
- add explicit constraints;
- add your own dimensions to get the scheme you want;
- modify dimension values to get desired size.

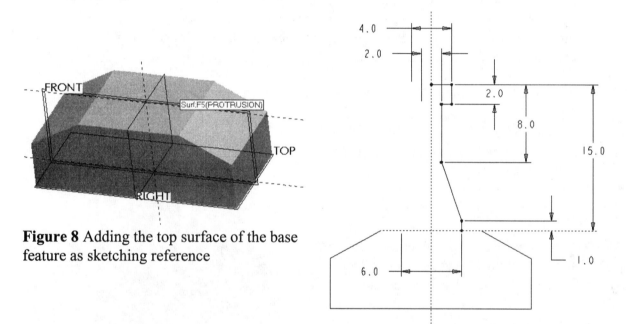

Figure 8 Adding the top surface of the base feature as sketching reference

Figure 9 Sketch for revolved protrusion

Here are a few tips for using Sketcher effectively:

1. While you are sketching lines, if a constraint appears in red, you can turn it off immediately (that is, prevent it from "sticking") by clicking the right mouse button.

2. The axis of revolution is specified using a centerline. To place a centerline along the vertical reference hold down the right mouse button in the sketch window and select *Centerline*. Click once on the vertical reference; the centerline will automatically snap to vertical when you click to create the second point on the reference.

3. Revolved features are usually specified using diameters rather than radii. To dimension a diameter of a revolved feature (the horizontal dimensions in the sketch): left click on the sketched line or vertex, then on the centerline, again on the same line/vertex, then middle click to place the dimension text.

4.   When you are modifying the dimension values, it is sometimes beneficial to do the smaller dimensions first. This ensures that the geometry will stay close to the desired shape throughout the changes.

Helpful Hint
If you have several construction lines in the sketch (they all have the same line style as a centerline), which one becomes the axis of revolution? You can choose the one you want by selecting it (highlights in red) then holding down the right mouse button to find the *Axis of Revolution* command. The axis of revolution will become yellow.

After finishing the sketch, select the *Accept* button. The most common error made here is forgetting to create the centerline used as the axis of rotation. Sketcher will catch this mistake and prompt you if you try to leave Sketcher without the centerline. You should get in the habit of always creating the revolve centerline first.

You should now be back in the Revolve dashboard. The default is a blind protrusion, which in this case means that the angle of the revolve is specified. It should be 360°. Since this is an open curve, Pro/E must know which side of the curve is to be made solid. That is the meaning of the arrow shown in Figure 10 (at the bottom of the revolve).

All elements should now be defined. *Verify* the part (it should look like Figure 10) and select *Accept*. Note that an axis has been defined as part of the feature. Although it doesn't appear in the model tree, the axis can still be used as another feature's reference (for example, a coaxial hole).

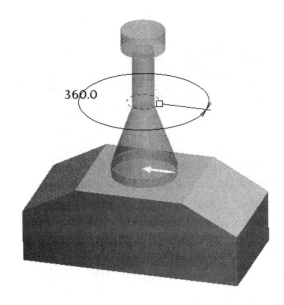

Figure 10 Preview of revolved protrusion

Adding a Pocket

We'll now use an extruded cut feature to create a pocket on one side of the base. Our design intent here will be to leave a 1 unit[1] thick wall around the pocket. We will use a useful tool in Sketcher to create this geometry. We have created a cut feature before.

Select the *Extrude* button in the right toolbar; its dashboard opens. Before we forget, click on the *Remove Material* button on the dashboard to produce the cut. Then, activate *Sketcher*. You should now be in the Section menu. Select the front surface of the block as the sketching plane,

---

[1] Incidentally, what are your units? These are the units of the default template. We'll introduce part units and how to change them in Lesson #8.

then select the top surface of the block as the TOP orientation reference. Middle click to enter Sketcher.

When you arrive in Sketcher, references will have been already picked. We will now create our sketch using only a single dimension - the thickness of the wall around the pocket!

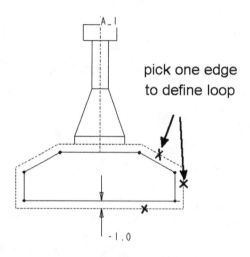

Figure 11 Picking edge to define loop

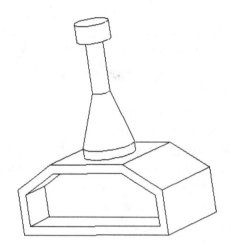

Figure 12 Pocket added using cut feature

On the flyout from the *Use Edge* button in Sketcher, select the *Offset* button ⬜ . In the **TYPE** window, select the *Loop* option. Pick on one of the right or bottom edges as shown in Figure 11, or you can pick on the front surface of the block. A small red arrow will appear on one of the edges showing an offset direction. Read the message window. If the arrow is pointing outwards, enter an offset value of *-1*, otherwise enter *1*. The sketch for our pocket is now complete, as in Figure 11. Notice the symbols on each sketched line, which indicates it was produced by an offset. Select the *Accept* button.

The Pro/E default is a blind cut, with the material removal side on the interior of our sketch. All we have to do is set the depth. Enter a value of **4** and *Accept* the feature. The resulting pocket should look like Figure 12.

Creating a Mirror Copy

(NOTE: As mentioned in the Introduction, this Tutorial was based on the pre-production release of Wildfire, Build 2002380. In this release, for functions described in this section, the program uses several pre-Wildfire menus and command sequences. It is possible that these may be superceded in a later build.)

Since the part is symmetrical, we can easily create the pocket on the back of the base by mirroring the first one. For the following, turn off the display of the datum planes (we are going to use a different tool to select these when required).

Starting in the pull-down menus, select

> ***Edit > Feature Operations***

Now we see some of the old pre-Wildfire menus. In the window that opens at the right, select

> ***Copy > Mirror | Select | Dependent | Done***

Figure 13 The Search window

The ***Mirror*** option is pretty obvious. Beware of the ***All Features*** option - it will select every feature in the part, including datums[2]. For the mirror operation, A ***Dependent*** copy means that if we change the geometry of the first pocket, the mirrored pocket will automatically be changed too. Use preselection to select and highlight the cut - the entire pocket should turn red. Then select ***OK > Done***.

See the message window. We want to mirror this pocket through the **FRONT** datum plane, which is currently turned off. To select this plane, we'll do something a little different. In the top toolbar, select the ***Search*** button 🔍.

The **Search Tool** dialog window opens (Figure 13). Here you can select references by datum or surface, and by name, ID, or feature number, and many other variations. The ***Look For*** data field has already been selected for us. Press the ***Find Now*** button. All the datum planes in the model are now listed at the bottom. In this list, select **FRONT** (it shows on the model) and ***OK***. The result is shown in Figure 14.

The ***Search*** command is handy if, as in this case, the feature is not displayed or if the model becomes very complicated with many datum planes and/or features. It is also helpful if the features are all named.

In the **FEATURE** menu at the right, select ***Done***. Or, middle click a couple of times to back out of this menu.

By the way, have you saved the part recently?

---

[2] Do not use the ***All Features*** option if you want a mirror copy of an entire part. There is a much more elegant way of doing this using a dummy assembly.

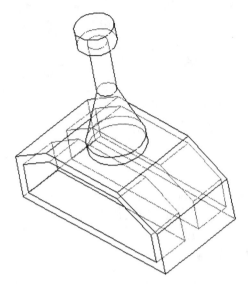

Figure 14 Part with mirrored pocket

Creating Holes

We already came across the hole feature in Lesson #3. We are going to add four holes as shown in Figure 1. We are going to do something a little different here with the depth specification, plus use the mirror command a couple of times after creating the first hole. For something a bit different, starting in the pull-down menu, select

> *Insert > Hole*

The **Hole** dashboard opens. The default is a *Straight* hole. For the primary reference (the placement plane), click on the sloping surface of the base at approximately the position where we want the hole center to be. This is shown ("placement plane") in Figure 15. Use the hole drag handles to select the **FRONT** datum plane (reference #1) and the upper edge of the end surface of the base (reference #2) for the linear references. This is one time where we must use an edge as a dimensioning reference, which we normally want to avoid. (Why?) The distance from each reference will be *3*. Set the hole diameter to *2.0*.

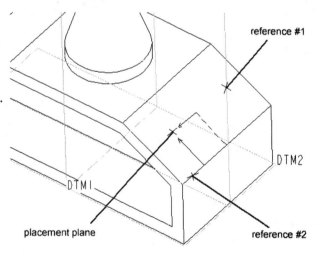

Figure 15 Placement plane and linear references

In the **Depth Spec** list, select *To Next*. As might be expected, this creates the hole until it passes through the next surface it comes to, wherever that is. The only restriction on *To Next* is that the sketch or hole must be entirely within the terminating surface. That is, if only part of the sketch or hole intersects the surface, the feature will just keep going through! We will see some examples of the problems this might cause when we get to the last section of this lesson.

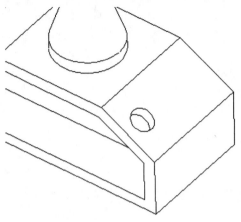

In this part, *To Next* means that the hole doesn't go all the way through the part. A blind hole may have achieved the same geometry, but would not be in keeping with our design intent. Why?

Figure 16 First hole

You can now *Verify* the hole. Assuming all is well, *Accept* the feature. See Figure 16.

We can use the mirror command to make copies of the hole. For practice, turn off the datum planes and use the *Search* command. The command sequence is

> *Edit > Feature Operations > Copy*
> *Mirror | Select | Dependent | Done*
> (select the hole)
> *OK > Done* (or middle click twice)
> (select the **FRONT** datum using *Search*)

The new mirrored hole should appear. We'll repeat this process to mirror both holes to the left side of the part at the same time using. The **FEATURE** menu should still be open, so select

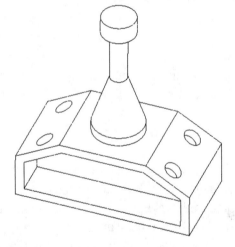

> *Copy*
> *Mirror | Select | Dependent | Done*
> (pick on the both holes, use CTRL-click)
> *OK > Done* (or middle click twice)
> (select the **RIGHT** datum using *Search*)

Figure 17 Holes added to base using *Mirror*

The part should now look like Figure 17. The figure does not show the axes created with each hole.

Having Problems Mirroring?

If you have trouble creating mirrored features, it is likely that your underlying geometry is not perfectly symmetrical about the mirror plane. We should not have that problem here, because we used the symmetry constraints on our base sketch, and a both-sides blind protrusion, which is also automatically symmetric. If you ever do have problems, you may have to double check the geometry to ensure that its dimensions are exactly correct. We will investigate this problem later on in this lesson. Geometric conditions at the location of the mirrored featured must be "legal" for the creation of the feature. For example, if the left side of the block did not have the same slope as the right side at the location of the hole, we should expect problems trying to do the mirror operation from right to left (since the hole must be perpendicular to the surface).

Creating Rounds (the *FAST* way!)

We will use a very handy short cut to add a couple of simple rounds to the top of the guide pin, and the edge where the shaft meets the base. Technically, these are called a round and a fillet, respectively. (A round removes material from an edge, while a fillet adds material.)

For the first round, use preselection to pick the edge where the base of the revolved protrusion meets the block (Figure 18). Note that only half the circular edge needs to be chosen - the feature will follow the tangent edge all the way around. The selected edge is highlighted in red. Now hold down the right mouse button, and select **Round Edges** in the pop-up menu. The display will show a preview of the fillet in yellow. Set the radius to **0.5** by entering the value, or using the drag handles. To accept the fillet, just middle click. That's fast! (How many mouse clicks?)

pick edge

Figure 18 Creating fillet at base

Figure 19 Preview of fillet at base

Let's do the same for the edges at the top of the shaft. Use preselection once again to highlight a single edge. Then use CTRL-click to select the second edge (see Figure 20). Once again, hold down the right mouse button and select **Round Edges** in the pop-up menu. Both edges will preview (Figure 21). The radius will be set initially to the value we used for the previous round.

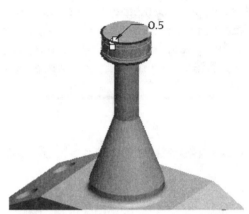

Figure 20 Selecting edges at top   **Figure 21** Preview of rounds at top

If you look in the **Sets** slide-up panel in the dashboard, you will see one set listed, containing two edges. All edges in the set have the same properties.

This round feature is OK for now, so just middle click.

If the tangent edges are not visible, select

> *Utilities > Environment*
> *Tangent Edges > Dimmed*
> *OK*

You might experiment with the image (shading, hidden line, no hidden, etc.) to see what the rounds look like in different displays.

Using Edge Sets

The last feature addition to this part is a chamfer all around the edge of the pocket and the parallel outside edge on the front and back of the base. We will include all these edges in a single chamfer feature by organizing them in two edge sets. All edges within each set have the same size. Edge sets are also used in the round feature. The main trick with edge sets is making sure a chosen edge is in the set you want it to be. Doing this is largely a matter of being careful when you are selecting the edges, and watching the screen carefully. The reason edge sets are useful is that the model tree can be simplified considerably by having multiple chamfered (or rounded) edges contained in the same feature. So, all chamfers or rounds of the same size can be modified simultaneously with a single dimension. Furthermore, when rounds or chamfers meet at corners, you can control the transitions between them. This can only be done for chamfers or rounds contained in the same feature.

You might find the following easier to do in hidden line or wireframe display. In shaded mode, some of the edges may not be really clear. Start by selecting the *Chamfer* command in the right toolbar. The dashboard is now open. You can immediately start selecting edges. Hold down the CTRL key and pick the six edges shown in Figure 22 below. The edges will highlight in red and the chamfer will show in preview yellow. When the last one is picked, adjust the size to **0.25** for this edge set. If you accidentally select a wrong edge, just pick it again.

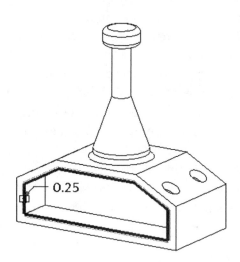

Figure 22 Chamfer edge set #1

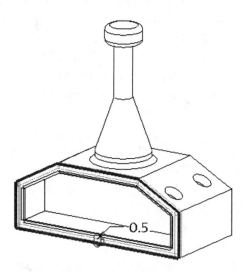

Figure 23 Chamfer edge set #2

Now, left click on an edge going around the outside of the base - see Figure 23. **As soon as you left click (without holding down CTRL), Pro/E assumes you are starting a new edge set.** The previous set is still shown in preview. Now holding down the CTRL key you can continue to pick edges for the second set. Set the dimension for this set to **0.5**.

Open up the **Sets** slide-up panel. The two sets are listed, with the edges of the highlighted set shown in the pane below.

There are no transitions here, since none of the edges intersect.

Verify the feature, and *Accept* it. See Figure 24.

Open the model tree and observe that the last feature is the chamfer, containing all 12 edges. The previous two features were the rounds. The mirrored holes appear in groups, as does the mirrored pocket. We will be discussing groups in a later lesson.

Figure 24 Chamfer completed

Now, in preparation for what we are going to do later (some things to make Pro/E fail!), we will try to mirror the chamfer to the edges of the pocket on the back face of the base using mirror plane **FRONT**. This seems like a reasonable kind of thing to do, but is kind of like poking the sleeping giant. Use the sequence starting in the pull-down menus:

> *Edit > Feature Operations*
> *Copy > Mirror | Select | Dependent | Done*

When you select the chamfer by picking on any of the chamfered surfaces, they will all highlight, since they all belong to the same feature. When it comes to specifying the mirror plane, select **FRONT**. Now the problems start! Pro/E is unable to create the mirrored feature and a **Failure Diagnostics Window** opens up. Click on the *<Overview>* field. This tells us how we can get some more information about fixing this problem. *Close* the information window. In the RESOLVE FEAT menu at the right, we have seen the *Undo Changes* command before - this comes up if a dimensional change has led to a regeneration failure. The problem here is different (and *Undo Changes* is not available anyway). To find out some more, select *Investigate > Show Ref*. This brings up the **Reference Information** window that we have seen before. In the list of parents, expand the list for the cut (click on the + sign). This brings up a list of all the references of the failed feature - the edges of the pocket. If you right click on any listed edge and select *Entity Info*, you will see a statement a few lines down to the effect that the edge is "not in the geometry." Hmmm ... it appears that Pro/E just won't let us do this mirror operation. We need to back out of this command. Close the **Reference Information** window. In the **RESOLVE FEAT** menu, select *Quick Fix > Delete*. This doesn't delete the original chamfer, just our attempted copy (the failed feature).

To get the chamfer on the back pocket we have two options:

1. Delete the existing chamfer and create a new one containing edge sets with edges on both front and back surfaces, or
2. Redefine the existing chamfer by adding new edges to the feature (this involves commands discussed in Lesson #5).

For now, you might as well try the first of these two. With the new chamfer with edges on both sides, the part is completed.

Saving the Part

Don't forget to save your part:

> *File > Save*

(or use CTRL-S) and if you have been saving regularly, get rid of previous copies of the part file by using

> *File > Delete > Old Versions > [guide_pin]*

and press the enter key (or middle click).

Model Analysis Tools

Quite often, you need to find some information about a part - its geometry, distances between points, surface area, center of gravity, moments of inertia. There are lots of tools available in Pro/E to query the model. Let's start with something simple. In the pull-down menus, select:

Analysis > Measure

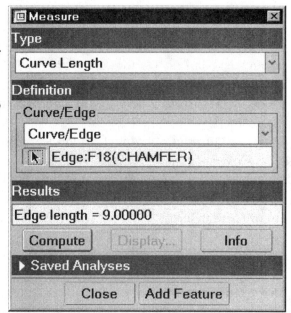

This opens the **Measure** dialog window (Figure 25). The default is to measure curve length. If you select any edge on the model, its length will appear in the Measure window and in the message area at the bottom. Select an edge of the hole. Is the reported length the full circumference or only half? What does this tell you about how Pro/E stores circular shapes?

In the pull-down list under *Type*, you will find options for angle, diameter, area, and so on. For example, select *Diameter*. Then pick on the cylindrical portion near the top of the shaft. The diameter shows in the Measure window. Even more interesting is to pick on the conical part of the shaft. This reports the diameter at the pick point.

Figure 25 Measure dialog window

If you select *Type(Area)*, you can select individual planar or curved surfaces (they highlight in pink) or the entire model. See Figure 26. This might be useful to calculate paint quantities. You can also obtained projected area by specifying a projection plane or direction.

By selecting *Type(Angle)*, you can find the angle between the sloping edges at each end of the block (Figure 27). *Type(Distance)* lets you find the distance between any two entities. Figure 28 shows the measurement of the (shortest) distance between the tangent edge of the round and the chamfer on the front of the part. You can also obtained projected distance.

The *Type(Transform)* measurement will produce a 4X4 homogeneous transformation matrix which describes the translation and rotation of one coordinate system relative to another.

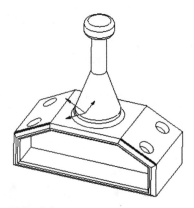

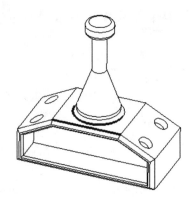

Figure 26 Area of surface

Figure 27 Angle between edges

Figure 28 Distance between curves

Close the **Measure** window. Now select (in the pull-down menus)

Analysis > Model Analysis

This opens the window shown in Figure 29. The default analysis is ***Model Mass Properties***. Press the ***Compute*** button at the bottom.

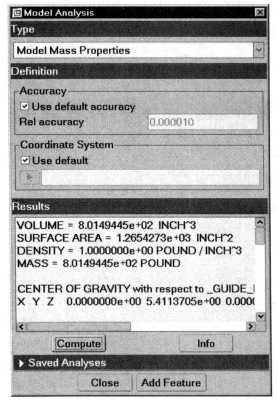

Figure 29 Model Analysis window showing part mass properties

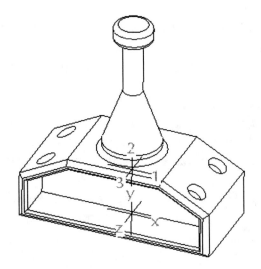

Figure 30 Part center of gravity and principle axes

The mass property data uses an assumed density for the solid (which can be changed). The results include the volume of the part, surface area, and mass. A bit farther down in the results is the location of the part center of gravity (relative to a specified coordinate system). The location of the center of gravity is shown graphically on the object (Figure 30). The axes 1-2-3 refer to the principle axes of the solid. The results also include the mass moments of inertia about these axes, and the radius of gyration of the solid about each axis. This data is useful for dynamic analysis of the part.

Other types of model analysis are cross section properties, one-sided volume (on one side of a selected plane), clearance between entities, determining if any edges are shorter than a specified length, or any thickness values are greater or less than specified values (important to molded plastic parts).

Exploring the Model, or "What Can Go Wrong?"

Now comes the fun stuff! Here are some things you can try with this part. These explorations are placed here rather than at the end of the lesson because they are very important. So DO NOT SKIP THIS SECTION!! We will review some of what we covered in Lessons #2 and #3. More importantly, some things we'll try will show you how Pro/E responds to common modeling errors such as the failure to mirror the chamfer earlier in this lesson. Being comfortable with these methods to respond when an error occurs is an important aspect of your modeling proficiency. A common tendency among newcomers to Pro/E is to retreat from these errors and try to create the model in another (usually more familiar but less efficient) way. This will not expand your knowledge of modeling practice, or allow you to anticipate errors before they happen. Spend the time to learn this now, and it will save you much time later.

Here are the exercises:

1.   We found out before how to name the features of a part. Do that now for the *guide_pin*, using whatever names you like. Obtain a feature list using
            Info > Feature List
      or
            Info > Model
      What is the difference between these lists?
2.   Make the following dimensional changes to various features of the model. (HINT: preselect the feature, use RMB pop-up menu and select *Edit*). *Regenerate* the part after making each dimensional change. Observe what Pro/E does and see if you can explain why. You can usually recover from any errors that might occur by selecting *Undo Changes*, or *Quick Fix > Delete*. If things really go wrong, you should be able to use *File > Erase > Current*, and retrieve your stored copy of the part file.
      ▸   change the radius of the round on the base of the revolved protrusion to the following values: (*0.75, 1.5, 3.0*). For each value, see if you can predict what Pro/E will do before you actually execute the regenerate command. Reset to the initial value (0.5) after these modifications.
      ▸   change the diameter of the first hole to the following values: (*1.0, 4.0*). Again, try to

predict how Pro/E will handle these changes. Reset to the initial value after these modifications. Try changing the diameter of one of the mirrored holes on the back of the part. When you click on this hole, where do the placement dimensions show up on the screen?

▸ change the location of the first hole from 3 to **1.5** away from the datum plane **FRONT**. Where does the hole now terminate? Why? Now change the same dimension to **5**. What happens and why? Reset to the initial value after these modifications.

▸ change the location of one of the holes from 3 to **1.0** away from the edge reference on the end of the block. Where does the hole now terminate? Now change the same dimension to (**7.0, 8.0**). What happens? Reset to the initial value after these modifications.

▸ change the height of the base block from 8.0 to *6.0*, then to *4.0*, then *3.0*. Explain what happens and reset to the initial value after these modifications.

▸ change the depth of the base block (10.0) to (*9.0, 8.25, 8.0*). What happens each time? Reset to the initial value after these modifications.

▸ change the length of the base block to (*16.0, 12.0*). Shade the view. What happens each time? Reset to the initial value after these modifications.

▸ change the radius of the base of the revolved protrusion (6.0) to the following values: (*8.0, 9.0, 9.5*). What happens? Reset to the initial value after these modifications.

▸ change the radius of the rounds on the top of the revolved protrusion to the following: (*0.75, 1.5*). What happens? Reset to the initial value after these modifications.

▸ change the edge offset dimension for the pocket to the following: (*2.0, 3.5*). What happens? Reset to the initial value after these modifications.

▸ change the depth dimension for the pocket to the following: (*4.5, 5.5*). Reset to the initial value after these modifications.

3.   Set up a relation so that the distance of the holes from the datum *FRONT* is such that the hole is always centered on the depth of the pocket. Add another relation that will give a warning if the web between the two pockets down the center of the part becomes less than 1.50 thick. Your relations will look something like this (your dimensions symbols will probably be different from these):

```
/* hole centered on pocket depth
d38 = (d5 - d14) / 2
/* narrow web warning - message is generated if false
(d5 - 2*d14) > 1.5
```

Check these relations by changing the depth of the base feature from 10 to 20. Then change the depth of the pocket to 9.5. Follow the prompts in the message window. When the part is regenerated, open the **Relations** dialog window.

4.   Examine the parent/child relations in the model. What are the parents of the pocket? What are the children of the pocket? Do the relations added in question 3 change the parent/child relations?

5.   Delete the front pocket and all its children. Now, try to create it again. What happens to the holes? Since this new feature will be added after the holes, you might anticipate some changes in the model. This points out again the importance of feature creation order.

6.   Explain why centering the base feature (the block) on the datums was a good idea.

7.   Try to delete the revolved protrusion. What happens?

8.   Try to delete one of the holes. What happens?

That's a lot of exercises and is enough to think about for this lesson. Select *File > Exit*. When you quit Pro/E, you might also have to check out your disk space usage and delete any files that you don't want to keep (for example: trail.txt).

In the next lesson we will discuss Pro/E utilities for dealing with features, including examining parent/child relations in detail, suppressing and resuming features, editing feature definitions, and changing the regeneration order. These are often necessary when creating a complex model, and to recover from modeling errors or poor model planning.

Questions for Review

1.   When sketching with Intent Manager, why should you deal with and set up your constraints before setting up the dimensioning scheme? Why do you set the dimension values last?
2.   What surfaces can be legally chosen as sketching planes?
3.   In Sketcher, how do you easily create an arc tangent to a line at an endpoint?
4.   What does the *Thru Next* depth specification do? What is a requirement for this?
5.   In Sketcher, where are the *Trim* and *Extend* commands? What do they do?
6.   What elements are required to create a **revolved protrusion**?
7.   What is meant by a **linear** hole? What are the alternatives?
8.   What is meant by a **dependent** copy?
9.   What is the difference between a round and a fillet?
10. What types of chamfer are available?
11. When you are creating a mirrored copy can you:
     ▸   select more than one feature to mirror at once?
     ▸   select more than one mirror plane at the same time?
12. What happens when a chamfer meets a round at the corner of a part?
13. What happens when two rounds of different radii meet at a corner of a part?
14. The figure at the right shows a sketch of two four-sided polygons. What is the difference between these polygons? Notice the difference in format of the "width" dimension, and the appearance of the vertical line on the far right.
15. Where do the placement dimensions of a mirrored feature appear under *Edit*?
16. What are the options for setting the depth of a blind, both-sides protrusion?
17. Could the rounds we made on the top of the guide pin be created as part of the **revolved protrusion**? What advantages/disadvantages would there be?

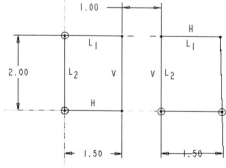

Figure for Question 14

Exercises

Here are some simple parts to make that use the features introduced in this lesson.

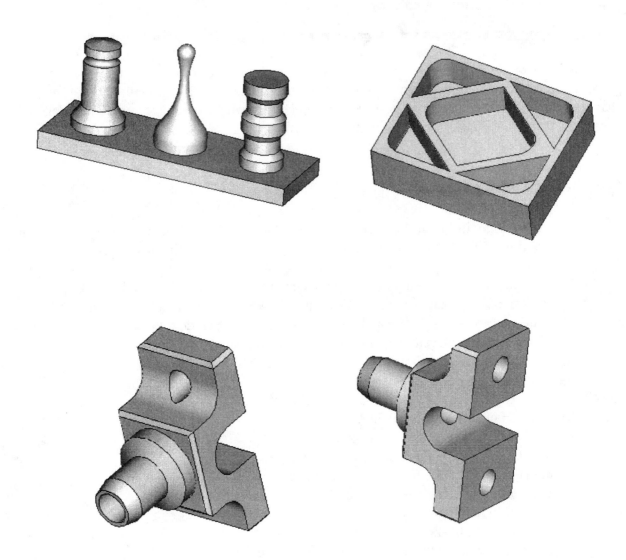

There are some more parts on the next page.....

These parts will be a bit more challenging. HINT: at this stage, keep your features as simple as possible, and plan ahead!

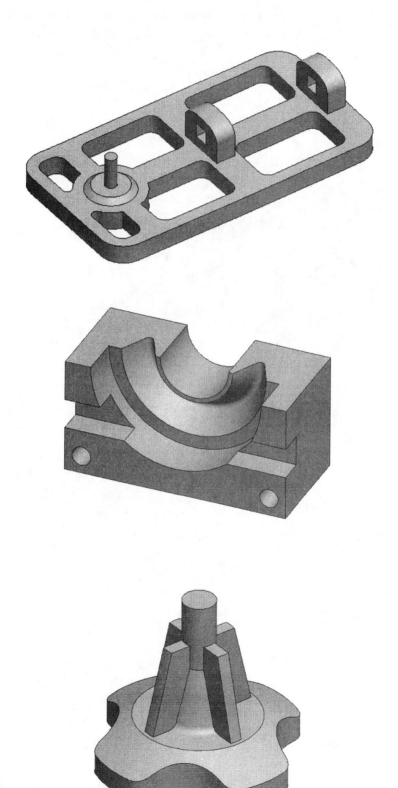

Project

Here is another part for the vise project, using features introduced in this lesson (a revolved protrusion, some mirrored cuts, and some rounds). All units are in millimeters.

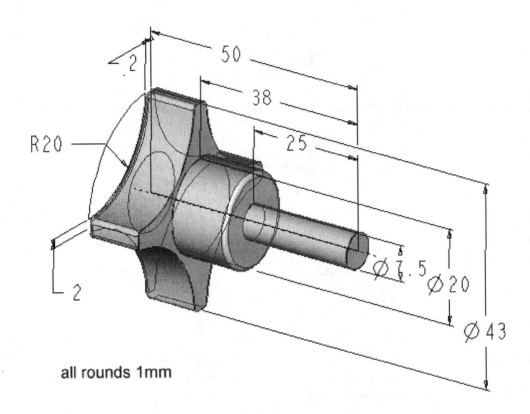

all rounds 1mm

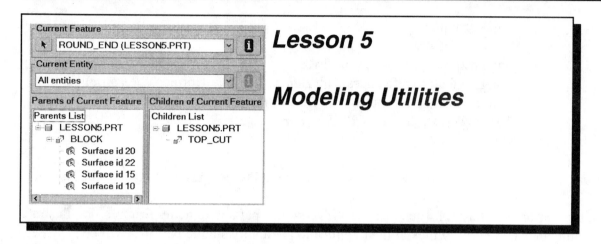

Synopsis

Utilities for exploring and editing the model: finding relationships between features, changing references, changing feature shapes, changing the order of feature regeneration, changing feature attributes, changing the insertion point, suppressing and resuming features.

Overview of this Lesson

In this lesson we will discuss Pro/E utilities for dealing with existing features. When you are modeling with Pro/ENGINEER, it is almost inevitable that you will have to change the geometry and/or structure of your model at some point. If your model becomes even moderately complex, you will need to know how to modify the model data structure and/or recover from poor model planning. This could be because you discover a better or more convenient way to lay out the features, or the design of the part changes so that your model no longer captures the design intent as accurately or cleanly as you would like. Sometimes, you just run into difficulty trying to modify the model (usually caused by the logical structure of the features) or have made errors in creating the model. This lesson will cover ways of obtaining information about parent/child relations, suppressing and resuming features, editing feature definitions and references (the commands formerly known as the 3 R's - **Redefine**, **Reroute**, and **Reorder**). We will also introduce Insert Mode for adding new features to the model early in the regeneration sequence. We have seen some of this before, so it will let you review that material.

The lesson is in four sections:

1. Obtaining Information about the Model
   - Regeneration Sequence
   - Obtaining a Feature List and Using the Model Tree
   - Getting Information about a Specific Feature
   - Parent/Child Relations
2. Suppressing and Resuming Features

- ▸ Single Features
- ▸ Handling Features with Children

3.   Modifying Feature Definitions - the 3 R's
- ▸ Changing feature references with ***Edit References*** (formerly ***Reroute***)
- ▸ Changing feature attributes with ***Edit Definition*** (formerly ***Redefine***)
- ▸ Changing creation order with ***Reorder***

4.   Insert Mode

As usual, there are Questions for Review, Exercises, and a Project part at the end of the lesson.

These utilities are most useful when dealing with complex parts with many features. To illustrate these commands here we will look at their application to a very simple part that will be provided for you. This part has a number of modeling "errors" that must be fixed. With parts this simple, it might actually be easier to just create a new part and start over again (you may find it necessary to do that occasionally anyway). However, when parts get more complex, and contain many features, starting over will not be an option and these utilities will be indispensable.

In order to do this lesson, you will need a copy of the file *lesson5.prt.1* that is available on the enclosed CD or from the SDC Web page (go to **http://www.schroff1.com**). Use your Web browser to download this file and copy it to your Pro/E working directory - full instructions to do this are on the Web page.

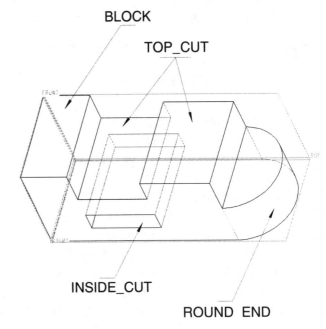

Once you have the part file, launch Pro/E, retrieve the part and continue on with the lesson. The part should look like Figure 1 in default orientation.

This model contains the default datum planes and four features. The base feature is a rectangular block. The other features are another solid protrusion and two cuts. The features are named as shown in Figure 1.

Figure 1 Initial part for Lesson #5

Obtaining Information about the Model

Once your model gets reasonably complex, or if you "inherit" a model from another source such as we are doing here, one of the important things to do is to have a clear idea of the structure of the model. Which features were created first? Which features depend on other features? How do the features reference each other? Answers to all these questions are available!

The Regeneration Sequence

The order of feature creation during part regeneration is called the *regeneration sequence*. Features are regenerated in the order in which they appear in the part database[1]. (We will talk about changing the order of the regeneration sequence in a later section of this lesson.) To observe the regeneration sequence select the following commands, starting in the pull-down menu:

Tools > Model Player

Enter a **1** in the **Feat #** box to go to the first feature. Then press the ***Step Forward*** button to step you through the creation of the model one feature at a time. The model player window will tell you which feature is currently being created. As you progress through the sequence, the menu gives you a chance to get more information about the current feature, including its dimensions.

For example, when you get to feature #6, select ***Feat Info*** in the **Model Player** window. This opens a Browser window with a page that gives you lots of information about the feature. Look for the following: *feature number* (#6), the *internal feature ID* (52), the ID's and feature numbers of the parents and children of this feature, the *feature type* (an extruded cut), dimensions. Note that the depth of this feature is indicated as *Variable* (this means Blind) with a depth of 10. This will be important later on. Also, note the difference between the feature number (the placement within the regeneration sequence) and the feature ID (Pro/E's internal bookkeeping). It will be possible to change the feature number, but, once created, you can never change a feature's ID.

Close the Browser window and continue through the regeneration sequence until you have all seven features. Then select ***Finish***.

The Feature List

You can call up a table summary of all the features in the model by selecting:

Info > Model

This brings up the Browser page shown at the right. At the top is shown the system of units for the model. Below this is a table that lists all its features. Information for each feature includes: the feature number and ID in the first two columns, a name for the feature (defaults to feature type), the type of feature, and current regeneration status. There are two action

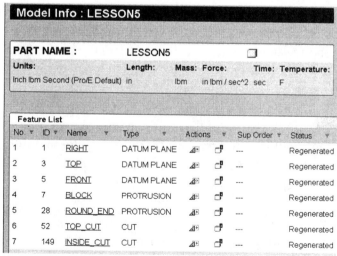

Figure 2 The Browser page for **Model Info**

---

[1]This is called "history-based" modeling. Some modeling programs do not have this restriction - they regenerate everything simultaneously.

buttons for each feature which will highlight the feature on the model, or open up the feature information page we saw previously. If you have many features, it is a good idea to name them - there is nothing worse than seeing a whole bunch of features all identified with just "Hole" or "Cut" in this table. Note that the feature names are links to additional information pages.

By the way, whenever you see a Browser window like this in Pro/E, you can easily print or save it by using the buttons at the top of the Browser window. This may be useful for design documentation. *Close* the Browser.

The Model Tree

The model tree was introduced earlier and you have probably seen it many times by now. If it is not currently displayed, open it by clicking on the textured button on the left sash. You should see the columns shown in Figure 3. If those are not visible, either load the model tree configuration file we made earlier (*tree.cfg*) or use

Settings > Tree Columns

to add and format columns. The usual columns you will use are *Feat #*, *Feat Type*, and *Status*. Also, while we're here, select

Settings > Tree Filters

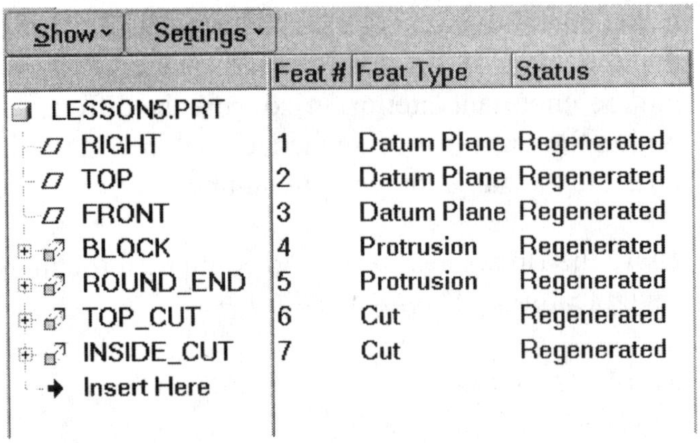

This brings up a dialog window with a number of checkboxes for selecting items to be displayed in the model tree. For example, remove the check mark beside **Datum Plane**, then select *Apply*. This might be useful if the part contains many

	Feat #	Feat Type	Status
☐ LESSON5.PRT			
▱ RIGHT	1	Datum Plane	Regenerated
▱ TOP	2	Datum Plane	Regenerated
▱ FRONT	3	Datum Plane	Regenerated
⊞ BLOCK	4	Protrusion	Regenerated
⊞ ROUND_END	5	Protrusion	Regenerated
⊞ TOP_CUT	6	Cut	Regenerated
⊞ INSIDE_CUT	7	Cut	Regenerated
→ Insert Here			

Show ˅   Settings ˅

Figure 3 Model tree with added columns

datum planes which are cluttering up the view of the model tree feature structure. Turn the datum plane display back on and exit the window with *OK*. (What happens if you *Close* this window instead?). The model tree should now look like Figure 3.

Left click on any of the feature names shown in the left column of the model tree to see it highlighted in the model. (If the feature doesn't highlight, make sure that **Highlight Model** is checked in the *Show* menu.) This is an easy way to explore the structure of the database and the features in the model. But the model tree can do much more!

Hold down the right mouse button on one of the features listed in the model tree. This pops-up a menu containing the following commands:

> ‣ Delete
> ‣ Suppress
> ‣ Rename
> ‣ Edit

▸ Edit Definition
▸ Edit References
▸ Pattern
▸ Setup Note
▸ Info

We have seen the *Edit* (for changing dimension values) and *Delete* (for removing features) commands before, as well as the *Info > Model* commands. The other commands *Edit Definition* (or *Redefine*), *Edit References* (or *Reroute*), and *Suppress* are among the main topics in this lesson, and are discussed at length below.

Parent/Child Relations

Using the commands given above, you can find out the regeneration sequence and internal ID numbers of parent and child features. There are several commands for exploring the parent/child relations in the model in considerably more detail. Select feature #5 in the model tree (ROUND_END) or preselect in the graphics window. Hold down the right mouse button and select:

Info > Parent/Child

The **Reference Information Window** (Figure 4) opens. The current feature is shown at the top of the window. Expand the lists in the Children and Parents areas.

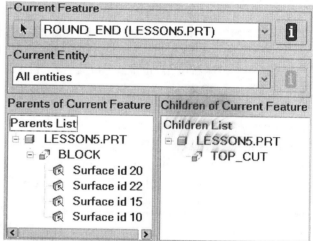

Figure 4 The Reference Information Window for feature ROUND_END

On the parents side, click on each of the four surfaces listed. As each is selected, the reference surface will highlight on the model and a brief message is given describing the nature of the reference (sketching plane or dimension reference).

On the children side, we see that the feature TOP_CUT is a child of the rounded end protrusion. What is the nature of this reference? Highlight this feature in the children list, then hold down the right mouse button and in the pop-up menu select *Set Current*.

The **Reference Information Window** now shows TOP_CUT as the current feature, and lists its parents and children. Expand these

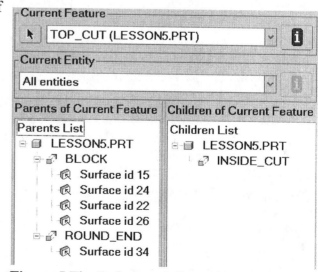

Figure 5 The Reference Information Window for feature TOP_CUT

lists (Figure 5). Notice the surface listed under ROUND_END in the parents area. Select this surface and it is highlighted on the model. The message tells us that this was used as the horizontal sketcher reference for the cut feature #6 (TOP_CUT). This will be important to us later. The other four parent surfaces of TOP_CUT correspond to the following:

1.   (Surface id 15) the front of the block - sketching plane
2.   (Surface id 24) top of block - dimensioning ref used for aligning/dimensioning the cut
3.   (Surface id 22) right end of block - dimensioning ref used for aligning/dimensioning the cut
4.   (Surface id 26) left surface of block - dimensioning ref used for dimensioning reference

If you repeat this process for the inside cut (use *Set Current* and expand the parent features), you should see the following references:

1.   the front of the block - sketching plane
2.   the right horizontal surface of the top cut - horizontal reference plane
3.   left vertical surface of the top cut - alignment/dimension reference
4.   right vertical surface of the top cut - alignment/dimension reference
5.   the Top datum plane - dimension reference

Now that we have explored the model a bit, you should have a good idea of how it was set up. Before we go on to ways that we can modify the model, let's have a look at a useful utility for dealing with features. Select *Close* in the **Reference Information Window** menu.

Suppressing and Resuming Features

When you are working with a very complex model, it will often happen that many of the model features are irrelevant to what you are currently doing. Or, you want to avoid accidentally picking on some features as a reference for a new one. There is a command available that will temporarily remove one or more features from the regeneration sequence (and hence the model display). This is called *suppressing* the feature(s). It is important to note that this does not mean deleting the feature(s), it just means that they are skipped over when Pro/E regenerates the model. This will speed up the regeneration process thus saving you time. On some systems, it may also noticeably speed up the screen refresh rate when doing 3D spins and shading.

When a feature is suppressed, it generally means that all its children will be suppressed as well. To bring the feature back, you can *resume* it. Let's see how suppress and resume work.

Preselect the feature INSIDE_CUT (or select it in the model tree). Hold down the

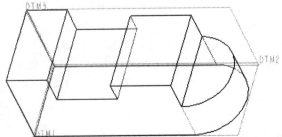

Figure 6 Part regenerated with cut suppressed

right mouse button, and in the pop-up menu select

> *Suppress*

Confirm the operation with **OK**. The part will regenerate without the cut as shown in Figure 6.

Check out the new feature list:

> *Info > Feature List*

Observe the feature status. Call up the model tree (make sure that ***Settings > Tree Filters > Suppressed Objects*** is checked). Notice the small black square beside the name of the suppressed feature.

You will note that the suppressed feature no longer has a feature number (but it still has an ID), and the last column shows its status as suppressed. To get the feature back into the geometry, issue the commands (starting in the pull-down menu)

> *Edit > Resume > Resume Last*

Now, try to suppress the TOP_CUT. Select it in the model tree, hold down the right mouse button and select **Suppress**. A warning window appears. Move it out of the way to see the model. The TOP_CUT is highlighted in red, the INSIDE_CUT is highlighted in green - it is a child of the TOP_CUT. You will have to decide what to do with it. Select **Options**. This opens the **Children Handling** dialog window (Figure 7). This window allows you to find information about the children (references and so on), as well as set options for how each child should be handled. The default action is to suppress all children with their parents. For now, select this with **OK** to suppress both cuts together. You should see the part as shown in Figure 8.

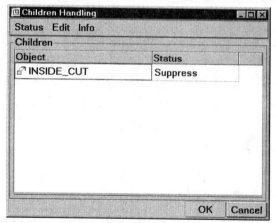

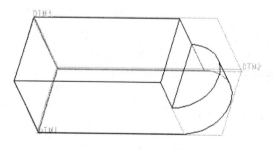

Figure 7 The Children Handling window   **Figure 8** Part with both cuts suppressed

Check the display in the model tree. Both features have the small black square indicating their suppressed status. Try to resume the INSIDE_CUT by itself. Select it in the model tree, hold down the right mouse button and select

> *Resume*

Both the selected feature and its parent (the other cut) are resumed - you can't resume a child without also resuming it's parent(s).

Using suppress and resume can make your life easier by eliminating unnecessary detail in a model when you don't need it. For example, if your part is a valve, you don't need all the bolt holes in the flange if you are working on some other unrelated features of the valve. If you are setting up a model for Finite Element Modeling (FEM) for stress analysis, for example, you would usually suppress all fine detail in the model (chamfers, rounds, etc.) in order to simplify it. Suppressing features also prevents you from inadvertently creating references to features that you don't want (like two axes that may coincide, but may be separated later). Finally, suppressing unneeded features will also speed up the regeneration of the part.

Features that are suppressed are still included in the part data base, and will be saved (with their suppress/resume status) with the part when you save your model to a disk file.

> **Helpful Hint**
> A trick used by advanced users who are dealing with very complicated parts is to suppress a large number of features before storing a part file. This reduces the file size, sometimes significantly. This can be useful when sending the part by email. When the file is read in again, the features are resumed by the new user. Be aware that this may be contrary to company policy (see next hint!).

> **Helpful Hint**
> When you inherit a part made by someone else, always check for suppressed features when you first open it. You may be (unpleasantly) surprised at what you find!

When we get to drawings and assemblies in the last lessons, remember that suppressed features are carried over into these objects as well. That is, a suppressed feature will stay suppressed when you add its part to an assembly, or display the part in a drawing. Suppressed features in a part may even prevent the assembly from regenerating since some important references may be missing.

Suppressing versus Hiding

Hold down the right mouse button on any of the datum planes. In the pop-up menu, you will see another command - *Hide*. Select that now. The datum plane disappears from the graphics window. All the datums in this part have children, so clearly we have not suppressed the datum, only removed it from the display. Observe that the datum icon in the model tree is now on a gray background. This indicates its hidden status. You can *Unhide* it using the same pop-up menu.

Hide and *Unhide* only work for non-solid objects (datum planes, curves, points). You cannot *Hide* a solid feature.

Modifying Feature Definitions

In previous lessons, we have used the *Edit* command to change dimension values. We need some tools to let us modify the basic structure of the model. So, now we will look at ways to modify the parent/child relations in the part, and to modify the geometric shape of some features.

Suppose we want to take the original *lesson5.prt* and modify it to form the part shown in Figure 9. This involves the following changes (some of these are not visible in the figure):

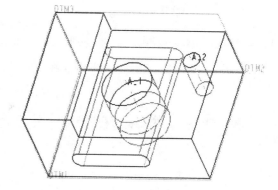

1.   delete the rounded end
2.   change the shape of the inner cut
3.   change the dimensioning scheme of the inner cut
4.   change the references of the inner cut
5.   change the shape of the cut on the top surface
6.   change the feature references of the top cut

Figure 9 Final modified part

7.   increase the depth of the part
8.   change the depth attribute of the top cut
9.   add a couple of vertical holes

Some of these changes will require modifications to the parent/child relations that were used when the part was created. This will also result in a cleaner model.

If you haven't gone through Section 1 of this lesson on obtaining model information, now is a good time to do so, since a good understanding of the existing parent/child relations is essential for what follows.

To see what we are up against, try to delete the rounded end of the part (the first thing on our "to do" list) by preselecting it, holding down the right mouse button and selecting

Delete

You will be notified that the feature has children (shown in green) and asked what you want to do with them. In the warning window, select *Options*. This opens the **Children Handling** window we saw before. In this window, select TOP-CUT. It will now highlight on the model. Hold down the right mouse button and select

Show References

You can now step through the references used to create TOP_CUT. As you step through these, they will highlight in green on the model. Watch the message window as you step through these. The first reference is the sketching plane. Select *Next* in the **SHOW REF** menu at the right. This highlights the horizontal sketching reference for TOP_CUT. The surface used was the upper surface of the rounded end - this is the parent/child connection that has interfered with our plan to delete the rounded end. We could change that reference now (using *Reroute*), but we'll deal with that possibility later. We could also delete the child along with the parent. We would then have to decide what to do with the children of the children (that is, the inside cut) and so on! Keep selecting *Next* in the **SHOW REF** menu to step through the rest of the references for TOP_CUT. When you have gone through them all, select

> *Done/Return*

in the **SHOW REF** menu. You are back to the **Children Handling** window. In the RMB pop-up menu for TOP_CUT, there are a couple of commands (*Replace References*, and *Redefine*) which we will discuss shortly.

In the **Children Handling** window, select the INSIDE_CUT and, in the RMB pop-up menu, select

> *Show References*

You will see the sketching surface, the sketching reference surfaces, and a couple of dimensioning references.

In the **Children Handling** window, the options for the two children are *Delete* or *Suspend*. The former will remove them from the model immediately (along with the parent). *Suspend* will keep them in the model, but the next time the model is regenerated, these features will fail regeneration and require special processing (like recreating or reassigning the necessary references that have been lost) to keep them in the model.

Cancel the deletion command that we launched previously. Clearly, this is not going to be as easy as it first looked. We'll deal with our desired changes one at a time, and not necessarily in the order given above. For example, before we can delete the rounded end, we have to do something about its child references. Some careful thought and planning is necessary here. When you get proficient with Pro/E, you will be able to manage these changes more efficiently. Our main tools to use here are: *Edit Definition, Edit References*, and *Reorder*. The first two commands were previously called *Redefine* and *Reroute*, respectively, and still appear with those names in a few places in Pro/E.

① Changing the shape of a sketch (*Edit Definition*)

The first thing we'll do is change the shape of the inner cut from its current rectangular shape to one with rounded ends. This requires a change in the sketch geometry of the feature. We'll take the opportunity to change the dimensioning scheme as well.

The ***Edit Definition*** (aka ***Redefine***) command allows you to change almost everything about a feature except its major type (you can't change an extrude into a revolve). Preselect the INNER_CUT (on the screen or in the model tree) and in the right mouse pop-up menu select ***Edit Definition***.

The feature dashboard will open, exactly as we saw it as the feature was being created. The feature is shown in preview yellow. The ***Remove Material*** button is selected and the **Depth Spec** is set to ***Through All***. Click on the button at the lower left to ***Activate Sketcher***, then in the **Section** window select ***Sketch***. Now we can use Sketcher to modify the sketched shape of the cut. You might like to go to wireframe or hidden line display here, and close the model tree. The desired final shape is shown in Figure 10.

First, CTRL-click on the vertical sketched lines at each end. With both lines highlighted, open the right mouse button pop-up menu and select

> ***Delete***

(or use the delete key on the keyboard). Now add two circular arcs: use the right mouse button pop-up menu to select

> ***3 Point / Tangent End***

and sketch the arcs at each end. Now change (if necessary) the dimensioning scheme to the one shown in Figure 10. Note that the ends of the straight part of the slot are still aligned with the vertical faces of the top cut. We will deal with those later. When the sketch is complete, select the ***Accept*** button. In the attribute window, select ***Verify*** if desired, then ***Accept*** the feature. If all went well, you should get the message

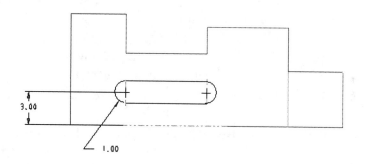

Figure 10 New sketch for the inner cut

> "Feature redefined successfully."

in the message area at the bottom of the screen.

② Changing a Feature Reference (*Edit References*)

Recall that the horizontal sketching reference for the inner cut was on the right side of the top cut, and we are planning on changing the shape of the top cut to remove that surface. We will have to change the reference for the inner cut to something else. This is done using the ***Edit References*** (aka ***Reroute***) command.

Preselect the inner cut, and use the pop-up menu to select *Edit References*. You will be asked if you want to *"roll back"* the model. Rolling back means temporarily returning to the part status when the inner cut was created. This is like suppressing all features created after the cut. This is a good idea, since then it will not be possible to (accidentally) select a new reference that is "younger" than the cut (ie. created after it). It is a good idea to **ALWAYS ROLL BACK THE PART!** It is curious that this is not the Pro/E default (although if you have a seriously complex model, this situation might change!) - you will have to enter a *y* (or click the *Yes* button) to cause the roll back to occur. This doesn't do anything for this simple model at this time because the cut was the last feature created.

The front surface of the block is highlighted in green on the part. The message window indicates that this was the sketching plane. The vertical and horizontal references of the sketch are shown. On the window right side, the **REROUTE REFS** window has opened. The command *Reroute Feat* is already active.

The general procedure in a reroute operation is to step through all the references for the feature being changed. In the **REROUTE** menu on the right, as you step through the sequence of current references, you have the options of selecting an alternate reference, keeping the same reference, or obtaining feature/reference information. As you step through the references, they will be highlighted on the part. Read the message in the message window - it will tell you what the currently highlighted reference is used for. For the inner cut, we want to:

1.   keep the same sketching plane (select *Same Ref*)
2.   select a different horizontal reference for Sketcher (select *Alternate*). A good one is the top surface of the block; an even better one is the horizontal datum plane (yellow side). Click on either of these now.
3.   keep all the same alignment and dimensioning references (select *Same Ref* three more times)

When you have cycled through all the references, you should get the message

"Feature rerouted successfully"

If you have rolled back the part, any features suppressed during the roll back will be resumed.

Go and check with *Info > Parent/Child* and click on the inside cut to confirm that the horizontal surface of the top cut is no longer referenced. There should still be a couple of references to the top cut, though. These are alignment constraints in the sketch of the inside cut. We'll have to change these if we are going to modify the top cut as planned.

③ **Changing the Sketcher Constraints (*Edit Definition*)**

As we saw earlier, the ends of the straight part of the inner cut are aligned with the vertical faces of the top cut. These alignments are still at work in the sketch. See Figure 11. To change these alignments, we need to redefine the sketch. So, preselect the INNER_CUT, and use the right mouse poop-up to select *Edit Definition*. In the feature dashboard, launch Sketcher.

Turn off the datum plane display. We want to do something with the sketch references so in the pull-down menu select

>   *Sketch > References*

Click on the left edge of the part. This should add an entry in the References window.   Now, select the other listed surface references (these will both be to feature #6, the top cut) and select the *Delete* button. The other two vertical references (shown in Figure 11) should disappear.

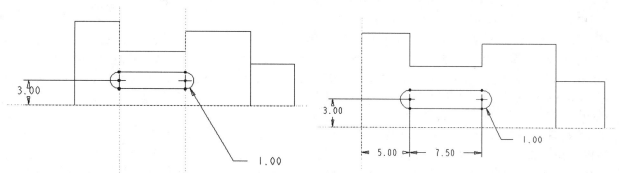

Figure 11 Old alignment references in the inside cut sketch

Figure 12 New dimensioning scheme for the inside cut sketch

Change the dimensioning scheme to the one shown in Figure 12. Intent Manager will do some of this for you automatically. Close out Sketcher and accept the redefined feature.

To make sure that there is now no relation between the top cut and the inside cut, preselect the TOP_CUT and, using the right mouse pop-up, select

>   *Info > Parent/Child*

The inner cut should no longer be listed as a child!

④ **Changing a Feature Reference (*Edit References*)**

Recall that the rounded end is a parent of the top cut via supplying the horizontal sketching reference. We need to break this connection before we can delete the rounded end (which is on our "to do" list). Pick on the top cut, use the right mouse pop-up, and select *Edit References*. Roll back the part. Notice that the inside cut disappears (temporarily). Keep the same sketching plane (*Same Ref*), but select a new horizontal reference (*Alternate*) like the top of the block or the horizontal datum. This is all we have to reroute, so select *Done*. You should get the message

>   "Feature rerouted successfully"

Select the rounded end and with

>   *Info > Parent/Child*

observe that it now has no children. Go ahead using the right mouse pop-up and selecting

> *Delete*

⑤ Changing Feature Attributes (*Edit Definition*)

We want to change the shape of the top cut to get rid of the step. We will also change its depth attribute. To see why this is necessary, select the BLOCK feature in the model tree, right click and select *Edit*. (Or just double click on the feature in the graphics window). Change the depth of the block from 10 to *15* and *Regenerate*. As you recall, the top cut had a blind depth of 10, so it doesn't go all the way through the new block as show at the right.

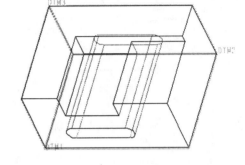

Let's change both the shape and depth of the top cut at the same time. Select it in the model tree, use the right mouse pop-up and select

Figure 13 Block width increased to 15

> *Edit Definition*

In the dashboard, change the **Depth Spec** option (or in the **Options** slide-up panel) from *Blind* to *Through All*. This fixes one problem. Now, to change the shape of the feature, select

> *Activate Sketch > Sketch*

Using the Sketcher tools, change the shape of the cut to a simple L-shape as shown in Figure 16. With the Intent Manager, you should be able to do this very quickly. Here are a couple of Sketcher tools to make this easier.

Select the *Dynamic Trim* tool ⬚. As the icon implies, hold down the left mouse button and drag the mouse cursor through the two lines you want to get rid of. See Figure 14.

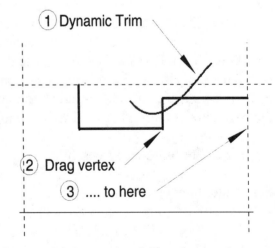

Now you can drag the right end of the horizontal line over to the right vertical reference of the sketch. To make sure the vertex sticks to this reference, open the Sketcher constraints menu and pick the *Same Point* constraint. Click on the vertex, then on the dashed reference line. You should now have the sketch shown in Figure 15. Make sure you dimensions match the figure.

Figure 14 A couple of Sketcher tools

When the sketch is complete, return to the feature dashboard.

Verify the part, and if it looks all right, select *OK*. The modified part is shown in Figure 16.

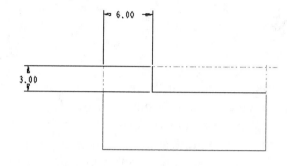

Figure 15 New sketch for the cut

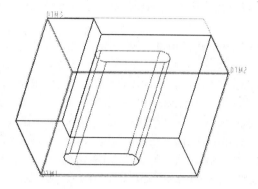

Figure 16 Part with redefined cut

⑥ Changing the Regeneration Sequence using *Reorder*

It is sometimes convenient or necessary to change the order of the features in the regeneration sequence. For example, an advanced technique involves grouping adjacent features in the regeneration sequence so that the group can be patterned or copied. The major restrictions on reordering features are:

- ♦ a child feature can never be placed before its parent(s)
- ♦ a parent feature can never be placed after any of its children

The reasons for these restrictions should be pretty self-evident. Fortunately, Pro/E is able to keep track of the parent/child relations and can tell you what the legal reordering positions are. To see how it works, in the pull-down menu select

Edit > Feature Operations > Reorder

and click on the inside cut, then *Done*. This cut (feature #6) was originally a child of the top cut (#5), but that relation was modified above. Thus, we should be able to create the cuts in any order, after the block (#4). In this simple part, there is only one legal possibility, that is, reorder the selected cut (currently #6) before the top cut (currently #5). This is what Pro/E tells you in the message window. In a more complicated part, Pro/E would tell you where the legal positions in the regeneration sequence are, and you could specify a *Before* or *After* placement for the reordered feature. Go ahead and complete the reorder: select *Confirm* and then call up the model tree. Note that the feature numbers of the cut and slot have now changed, but the internal ID's are still the same. Close the **FEAT** menu with *Done*.

Pro/E has made the ***Reorder*** command quite a bit easier by allowing you to drag and drop features in the model tree. Try that now by reordering the top cut. Click on the feature in the model tree and slowly drag the cursor upwards. The mouse icon changes slightly as you move back up the list to show you where legal reordered positions are. In this part, of course, there is only one valid position. You might try out this mode of reordering sometime when you get a more complicated part.

⑦ Changing the Insertion Point

New features are typically added at the end of the regeneration sequence (notice the "Insert Here" arrow in the model tree). Sometimes it is necessary to create a new feature whose order you want to be earlier in the sequence. You could do this by creating it and then using the reorder command, being careful that you don't set up parent references to features after the targeted reorder position. Also, you would have to be careful not to create any new features that could interfere with existing features (like cutting off a reference surface). There is an easier way!

In the pull-down menu, just select

> ***Edit > Feature Operations***
> ***Insert Mode > Activate***

You will be asked to select which feature to insert after. Pick on an original surface of the block (not one created by either of the cuts) or pick the block in the model tree. The part will automatically roll back by suppressing all features created after the block. Notice the new position of the "Insert Here" arrow in the model tree.

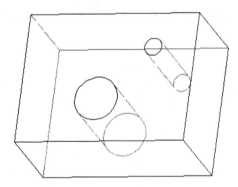

Create two *Through All* circular holes in the part as shown in the figure (the diameters are 2 and 5; placement is approximately as shown).

Figure 17 Two holes inserted after block feature

Insert mode will stay on until you turn it off by selecting

> ***Edit > Feature Operations > Insert Mode > Cancel***

You will be asked about resuming the features; accept the default [Y, or middle click]. Call up the **Feature List** or model tree to see that the two holes have been added to the model after the block and before the cuts.

You have probably guessed that you can also activate insert mode by dragging the ***Insert Here*** arrow back up in the model tree. Try it! You can move the insertion point around pretty much anywhere in the model tree. There is one place it won't go - can you find it? All the features after the insertion point are automatically suppressed. If you resume any of these features, the insertion point will advance to the last feature resumed in the model. This is a handy way to step through a model (much like using the model player).

Conclusion

The modeling utilities described in this lesson are indispensable when dealing with complex parts. You will invariably come across situations where you need to redefine, reroute, or reorder features. The information utilities are useful for digging out the existing parent/child relations, and discovering how features are referenced by other features. The more practice you get with these tools, the better you will be able to manage your models. As a side benefit, having a better understanding of how Pro/E organizes features will cause you to do more careful planning prior to creating the model, with fewer corrections to be made later. This will save you a lot of time!

In the next lesson, we will investigate the use of datum planes and axes, including creating temporary datums called "make datums". We'll also discover yet more tools and commands in Sketcher.

Questions for Review

1. How can you find out the order in which features were created? What is this called?
2. How can you find which are the parent features of a given feature?
3. How can you find the references used to create a feature?
4. How can you find any or all other features that use a given feature as a reference?
5. What is the difference between the Feature # and the internal ID?
6. What is the command to exclude a feature temporarily from the model?
7. What happens to the parents of a suppressed feature? To the children?
8. Is it possible, via a convoluted chain of parent/child relations, for a feature to reference itself?
9. What happens to suppressed features when the model is saved and you leave Pro/E?
10. If you are given a part file that you have never seen before, how can you determine if it contains any suppressed features?
11. In Sketcher, how many variations of the right mouse pop-up menu can you find? In what modes are these active?
12. How can you restore previously suppressed features?
13. How can you change the sketch references when you are in Sketcher?
14. How many features can you suppress at once?
15. Can you resume a parent without resuming its children?
16. Is there any aspect of a feature that cannot be modified using *Edit Definition*?
17. What is the difference between *Edit*, *Edit References*, *Edit Definition*? Which is the most

general command?
18.   What is meant by "rolling back the part?"
19.   How can you remove unwanted alignments in a sketch?
20.   What symbol in the model tree indicates suppressed features?
21.   What are the two fundamental rules of reordering?
22.   Are there any restrictions on the insertion point in *Insert Mode*?
23.   What happens if Insert Mode is on when you save a part and then later retrieve it?
24.   How do you get out of insert mode?

Exercises

Here are some simple parts to model using the features we have covered up to here. Before you
start creating these, think about where you will place them relative to the datum planes, what type
and order you should select for the features, and how you should set up parent/child references
and dimensioning schemes.

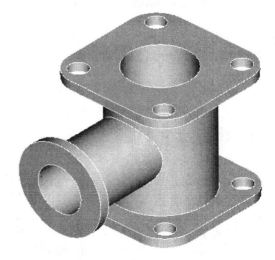

Project

Here are three small parts for the project. All dimensions in millimeters. For the acorn nut, you might like to investigate alternate Sketcher environments (see *Sketch > Options* when you are in Sketcher), including a polar grid, and the use of centerlines as construction aides (straight lines and/or circles). Note that the hexagon on the nut requires only one dimension to give its size.

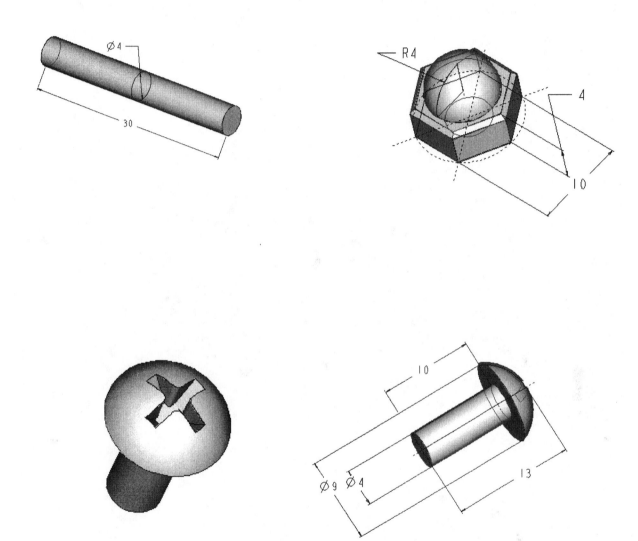

This page left blank.

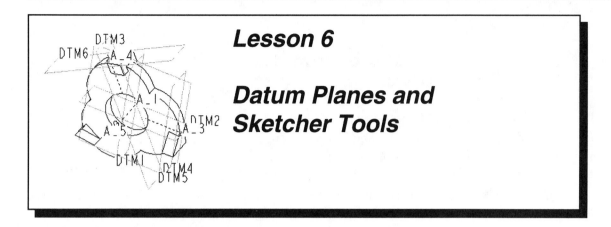

Lesson 6

Datum Planes and Sketcher Tools

Synopsis

The mysteries of datum planes and make datums are revealed! What are they, how are they created? How are they used to implement design intent? More tools in Sketcher are introduced.

Overview of this Lesson

In this lesson we are going to look at some new commands in Sketcher for creating sections. We will also use relations within Sketcher to control the geometry. Our primary objective, though, is to look at the commands used to set up and use datum planes. Some of these datum planes, like the default ones (**RIGHT**, **TOP**, and **FRONT**), will become references for many features, or will appear similarly on the model tree. Others, called *make datums*, are typically used only for a single feature and are created "on-the-fly" when needed. Along the way, we will discuss some model design issues and explore some options in feature creation that we have not seen before. The part we are going to create is shown in the figure at the right.

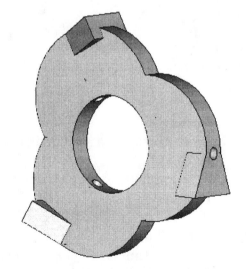

As you can see, the part consists of a three-lobed disk with a central hole. Three triangular teeth are spaced at 120 degrees around the circumference. Each tooth includes a central radial hole that aligns with the central axis of the disk. Although there is no indication of it in the figure, each of these tooth/hole features will be created differently using different datum plane setup procedures. We will see what effect this has on the model at the end of the lesson[1].

Figure 1 Final part - three tooth cutter

---

[1]A better way to create this part would be to create a single lobe, tooth, and hole, then group these together. The group can be copied around the central axis, creating a "pattern." We will have a look at patterns in the next lesson.

Here is what is planned for this lesson:

1.   Overview of Datum Planes and Axes
2.   Creating a Datum Plane and Datum Axis
3.   Create the Disk with Hole
4.   First Tooth - using an Offset constraint
5.   Second Tooth - using Normal and Tangent constraints
6.   Third Tooth - using Make Datums
7.   Effects on the Model
8.   Things to Consider about Design Intent

As usual there are some Questions for Review, Exercises, and a Project part at the end of the lesson.

Overview of Datum Planes and Axes

Datum features (planes, axes, curves, points and coordinate systems) are used to provide references for other features, like sketching planes, dimensioning references, view references, assembly references, and so on. Datum features are not physical (solid) parts of the model, but are used to aid in model creation. Datum planes or axes extend off to infinity. By default, Pro/E will show visible edges of a datum plane or the datum axis line so that they encompass the part being displayed. It is possible to scale a datum plane differently so that, for example, it will extend only over a single feature of a complex part. This would be done to reduce screen clutter. When we use the word "datum" by itself, we usually mean a datum plane.

Let's consider how a datum plane can be constructed. In order to locate the position and orientation of a datum plane, you will choose from a number of constraint options. These work alone or in combination to fully constrain the plane in space. The major options for datum planes are:

Through
        the datum passes through an existing surface, axis, edge, vertex, or cylinder axis
Normal
        the datum is perpendicular to a surface, axis, or other datum
Parallel
        the datum is parallel to another surface or plane
Offset (linear)
        the datum is parallel to another surface or plane and a specified distance away
Offset (rotation)
        the datum is at a specified angle from another plane or surface
Tangent
        the datum is tangent to a curved surface or edge

Some of these constraints are sufficient by themselves to define a new datum plane (for example, the **Offset(linear)** option). Other constraints must be used in combinations in order to fully constrain the new datum. When you are constructing a new datum, Pro/E will automatically pick an appropriate option (based on the entity selected) and show you the results of the currently set constraints by using a preview.

Construction of a datum axis is similar, with the following constraint options:

Through
> the axis is through a selected vertex, edge, or plane. May require addition constraints.

Normal
> the axis is located using linear dimensions and is normal to a selected plane

We won't have time to explore all the variations of these options in this lesson. The general procedure is pretty similar for all options, however. With the new preview capability in Wildfire it is quite easy to figure out what to do after you have seen the general procedure a few times.

Let's see how this all works. Start Pro/E in the usual way, and clear the session of any other parts. Start a new part called *cutter* using the default template. The default datum planes are created for you as the first features in the part. You can delete the datum coordinate system feature for this part since we won't need it and it just clutters up our view (or just turn off its display with *Hide*).

After the default datums are created, new datum planes and axes are created using either the toolbar buttons on the right of the graphics window (the ones without the eyeballs) or using the *Insert > Model Datum* pull-down menu at the top.

Creating a Datum Plane and Datum Axis

First, we will define a datum axis that will be the central axis of the cutter. This will be at the intersection of the existing datums **RIGHT** and **TOP**. Using the CTRL key, select both these datum planes. They should both be highlighted in red. Now, select the *Datum Axis* button in the right toolbar. This creates the datum axis A_1, as shown in Figure 2. This is the appropriate axis for Pro/E to make if you pick two intersecting plane surfaces, so it skips over the **Datum Axis** dialog window. Let's have a look at that window.

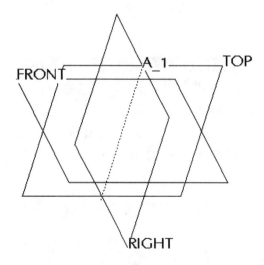

Figure 2 Datum axis A_1 created

With A_1 highlighted, use the RMB pop-up menu and select
Edit Definition. This opens the **Datum Axis** dialog shown in
Figure 3. We see the two datums, RIGHT and TOP, each with
the constraint ***Through***. This fully constrains the axis. Close
the window with OK.

In the procedure we just did, we selected the new datum
references first, then launched the datum creation command.
This is an example of object/action execution. Alternatively,
we could have launched the ***Datum Axis*** tool first to open the
dialog window, then picked on each of the datums and set the
associated constraints shown in Figure 3 to ***Through***. This
would be an action/object procedure. You can use whichever
method you are most comfortable with - they have the same

Figure 3 Datum Axis dialog
window

final effect on the model. Once you get more experience with the commands, you will probably
find object/action to be more efficient.

Our next task is to create a new datum plane that passes through the axis A_1 we just made, and
is at a specified angle to the RIGHT datum plane. We will use this as a reference in a couple of
features later on in the part.

We will use action/object for this. Select the ***Datum Plane*** tool in the right toolbar. The **Datum
Plane** dialog window opens at the right. We need to specify the references for the new datum.
Pick on the axis A_1. This is added to the reference list in the dialog window with a ***Through***
constraint. Click on the listed constraint to see a hidden pull-down constraint list; the only
alternative now is ***Normal***. Leave it set to ***Through***. Now, holding down the CTRL key, select
the RIGHT datum. This is now listed as well, with an ***Offset*** constraint. The ***Offset(rotation)***
constraint is the only one that makes sense with the existing ***Through*** constraint. The datum is
previewed on the graphics window, with a drag handle to control the value of the offset angle.
This value is also shown in the dialog window (Figure 4). Set the value to ***60***. A negative value
would rotate the other way. The datum plane preview should look like Figure 5 (datum plane in
yellow).

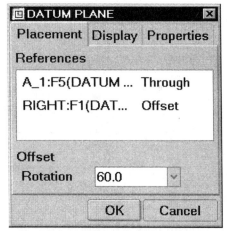

Figure 4 Datum Plane dialog
window

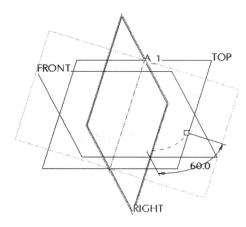

Figure 5 Datum plane with ***Through*** and
Offset(rotation) constraints

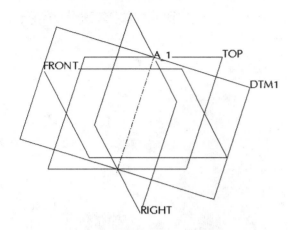

Figure 6 Datum plane DTM1 created at 60 degrees from RIGHT

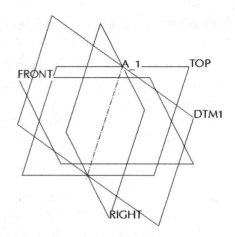

Figure 7 Datum plane DTM1 changed to 30 degrees from RIGHT

In the **Display** and **Properties** tabs of the dialog window you can change the normal direction (the yellow arrow) and name of the feature. Leave these alone and select ***OK***. The new datum plane DTM1 will be added to the model (Figure 6). If you open the model tree, you will see features A_1 and DTM1 there. If you double-click on the edge of DTM1, you will see the offset angle dimension. Change that to **30**, and ***Regenerate*** the part (Figure 7). This angle dimension is how we control the orientation of the datum plane.

These two examples have illustrated the general procedure for creating datum planes and axes. You might like to come back and experiment with these. There are some short cuts available that can save you some time (like preselecting features before launching the Datum command). To use the shortcuts, you need to have a very good grasp of how Pro/E will utilize defaults and the references you give it. You should also spend some time exploring the various constraint options for each chosen feature - these are available in a pull-down list beside each feature in the dialog window. Possibly because datums do not result in solid geometry, new users tend to find them a little tricky to deal with and, as a result, often do not make very effective use of them. Remember that we are creating a model, not just a solid. Datums are often crucial elements of the model structure.

We will leave datums for a bit now, so that we can create the base feature of the cutter. We will be creating several more datums as we go through this lesson.

Creating the *Cutter* Base Feature

Our base feature is a solid protrusion that will look like the figure shown at the right. We are going to go through the sketching procedure slowly here to illustrate a few new tools and techniques.

Our plan of attack is to sketch this shape on the FRONT datum plane. Since the part is symmetric front to back, we will make this a **Symmetric, Blind** protrusion. It also makes sense to center the feature where the datums TOP and RIGHT meet. The reason for DTM1 will be clear when we get into the sketch - it provides a reference for locating the geometry.

The sketch basically consists of three circular arcs. We'll call these the first, second, and third arcs, starting at the right and going counterclockwise.

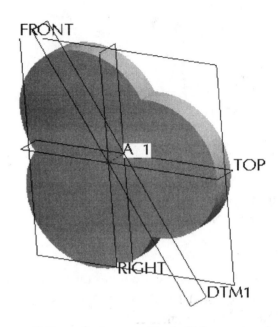

Figure 8 Base feature - a **Both Sides** protrusion

Select the **Extrude** tool. Its dashboard opens up. Change the **Depth Spec** to **Symmetric** (also known as **Both Sides**), and set its value to **2**. Note that we can do this even before we have created the sketch. Now select the **Activate Sketcher** button. Click on the FRONT datum as our sketching plane. The RIGHT datum becomes our orientation reference. This is fine, so middle click to enter Sketcher.

In Sketcher, two references have been chosen for us. We want to add DTM1 (which is visible on edge) as a reference, so select that and then **Close** the **References** window.

The center of each of the three arcs is the same distance away from the axis A_1. We can implement this intent in Sketcher by creating a construction circle. Use the RMB pop-up menu to select **Circle**, and draw the circle shown in Figure 9. Set the diameter to **8**. Click on the circle so that it highlights in red and open the RMB pop-up menu. Select the **Construction** command. This changes the line style to dashed. This curve can now be used as a sketching reference and will not contribute to solid geometry of the feature. You can toggle a construction line back to a physical edge using the same RMB pop-up menu command.

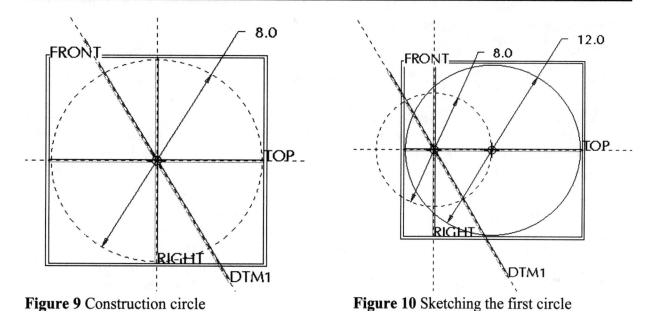

Figure 9 Construction circle   **Figure 10** Sketching the first circle

Now (use RMB pop-up) select the *Circle* command again and draw the first circle, shown in Figure 10. Its center can be snapped to the intersection of the construction circle and the reference on TOP. Set the new circle diameter to 12.

Select the *Circle* command again to draw the second circle. Its center can be snapped to the intersection of the construction circle and the reference on DTM1. Drag out the circle until the "equal radius" constraint snaps in. Look for a red R_1. Complete the circle - see Figure 11. You might turn the datum planes off now, since the screen is getting a bit cluttered.

Select the *Circle* command for a third time. The center we want is on the construction circle and directly below the center of the second circle. The cursor should snap to this position. Watch for the small blips that indicate vertical alignment. Once again, drag out the circle until the R_1 constraint snaps. The sketch should look like Figure 12. Weak constraints are shown in gray, so may be hard to see.

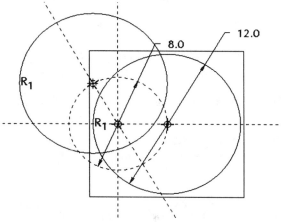

Figure 11 Sketching the second circle

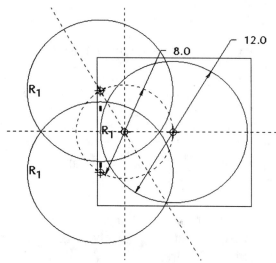

Figure 12 Sketching the third circle

Now we want to trim away all the geometry inside the three arcs. Here are a couple of tools to do that. First, there is a nifty tool on the ***Trim/Divide*** flyout that looks like this . As the icon implies, all you have to do is swipe the mouse pointer across the edge segments you want to remove - this is ***Dynamic Trim***. The edge will be trimmed back at both ends to the nearest intersection point or vertex. Try it! See Figure 13. When using this tool, one thing you will have to watch out for is the presence of very small line fragments left behind after trimming. You can usually spot these either by the blue dots on the vertices or by dimensions that seem to go nowhere. To get rid of all these fragments at once, you may have to resort to another trick for deleting entities. Make sure ***Select*** is picked in the Sketcher menu. Then left click and drag out a rectangle that encloses all the offending lines. They should highlight in red. If you want to remove something in this selection set (this might include constraints), use the CTRL key when you pick the item to toggle its selection status. For example, in this sketch we do not want to delete the construction circle, so remove it from the selection set. Hold down the right mouse button and select ***Delete***. The selected entities are all gone! You may have accidentally deleted some Sketcher constraints (like vertical alignment of the centers of the second and third circle). Intent Manager is able to generate other constraints to keep the sketch solved. The completed sketch should look like Figure 14.

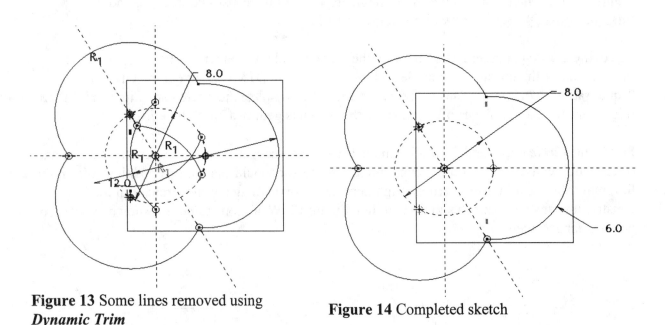

Figure 13 Some lines removed using ***Dynamic Trim***

Figure 14 Completed sketch

To test the flexibility of this sketch (and if the constraints are doing what we want), try changing either of the dimensions. For example, change the diameter of the construction circle to 6 and/or the radius of the arc to 8. Try some other values. If the sketch is robust, it should be able to regenerate correctly for a wide range of dimensional values. Return the dimensions to the values shown in Figure 14.

Accept the sketch. Back in the extrude dashboard, ***Verify*** the feature. The solid should be symmetric about FRONT (check the right side view). If everything looks like Figure 8 above, ***Accept*** the feature.

Creating a Coaxial Hole

We'll create the large center hole using some new options in the hole dialog window. Preselect the axis A_1. If you have trouble picking this, try setting the **Filter** at the bottom right to *Datums*. Now, with A_1 highlighted, select the *Hole* command in the right toolbar. A one-sided blind hole is now previewed. Change the diameter to *8.0*. Open the **Placement** slide-up panel (Figure 15). Since we entered the command with A_1 preselected, it has become the primary reference,

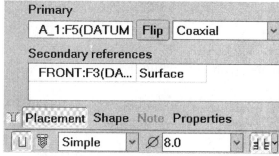

Figure 15 Creating a coaxial hole

and Pro/E assumes that we want a *Coaxial* hole. It only needs to know what surface the hole should be defined on - this is the secondary reference whose pane has no items. Click in this pane and then select the FRONT datum on the screen. The hole is still one-sided and blind. In the **Shape** slide-up panel, change the depth spec to *Through All* . In the same panel, change the Side 2 depth spec also to *Through All*. The preview now shows the hole coming in both directions off FRONT, with no depth dimension.

That completes the hole, so you can *Accept* the feature.

Have you saved the part yet? Now is a good time.

First Tooth - Offset Datum

The first tooth will be the one at the right (3 o'clock position). The design intent for this tooth is that the inner side of the tooth will be a specified distance away from the disk axis. We will create a datum plane at the desired distance that we can use as a sketching plane. The tooth will be extruded outward (for a fixed distance) to the outer edge of the disk. Then we will place a hole, also on the new datum plane, using the both sides option to go radially inward and outward.

Start by selecting the *Datum Plane* button on the right toolbar. Then pick the RIGHT datum plane. This reference will be listed in the dialog window on the right, with the default constraint *Offset*. Note that this is a translation. On the yellow preview of the new datum, you will see a single drag handle. Drag this out to the right. The dimension shows the offset distance from the reference. The offset dimension is also given in the dialog window.

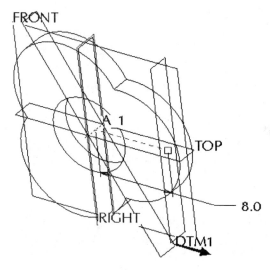

Figure 16 Creating an *Offset(linear)* datum

If you wanted to go to the other side of **RIGHT**, you could enter a negative offset. For now, enter a value of *8*. *Accept* the new datum. It will be called **DTM2**.

Now we can create the tooth. Select the *Extrude* command. In the dashboard, activate the Sketcher button. In the **Selection** window, we will select the new datum DTM2 as our sketch plane. The TOP datum will be automatically select for the top orientation reference.

In Sketcher, the two references TOP and FRONT have been chosen for us already. We want to add a couple more to this list. Spin the model a bit so that you can pick on the front, then the back, surfaces of the base feature. We prefer to use surfaces for references instead of edges. You can delete the reference FRONT.

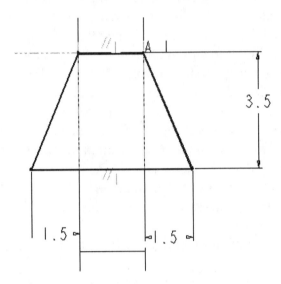

Make the sketch shown in Figure 17. Notice that the top line in the sketch aligns with the horizontal reference. This sketch implements a design intent where the width of the tooth is determined by the overhang beyond the side of the disk. Can you think of different ways of using references and dimensions to create different design intents for this sketch?

Figure 17 Sketch of first tooth

Let's add a relation to make sure the two overhangs are the same. We'll do this the fast way presented in a previous lesson. In the pull-down menus, select

> *Info > Switch Dimensions*

The *sdx* labels appear on the sketch. Note the label for the overhang on the right of the sketch. Now double-click on the overhang dimension on the left. You don't need to have the dimensions in numeric form. Enter the dimension symbol for the right overhang distance. You will be asked to confirm adding this relation to the sketch. Now, select

> *Info > Switch Dimensions*

to get back to the numeric display. As usual, when you have created relations you should test them to make sure they are working properly. Try to change the value on the left - you can't. Try changing the one on the right - they should both change. Return the value to the one shown in Figure 17.

When you have a completed sketch, leave Sketcher, select a *Blind* depth of *2*, and accept the feature. Turn the display of the datums back on.

Create the small hole using the new datum plane DTM2 as a placement surface. This is a straight, linear hole. Once you have selected DTM2 as the placement plane, drag the two linear placement handles to the FRONT and TOP datums (anywhere on the displayed edges of these datums will do for an attachment point). Both dimension values should set to *0*. If you have trouble picking these dimensions on the screen you can also set these in the **Placement** slide-up panel. The hole diameter is *1.0*. In the **Shape** slide-up panel, set both the Side 1 and Side 2 depths to *To Next*.

Note that in one direction, a *Through All* depth would have gone completely through the other side of the disk, which we don't want. *To Next* extends the hole until it passes through the next part surface. *Accept* the hole.

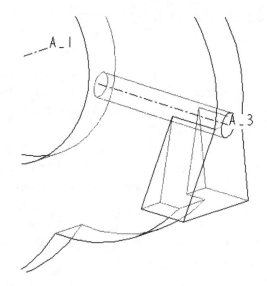

Figure 18 First tooth and hole complete

The tooth/hole combination should now be complete and look like Figure 18.

Second Tooth - Normal and Tangent Datum

The second tooth is the one at the top left of the part (on arc #2). The intent demonstrated here is to have the planar outer surface of the tooth tangent to the arc of the disk and to extrude the tooth inwards towards the center of the disk. So, we will create a datum to give us a flat sketching surface at the outer edge and tangent to the disk. We can make use of our existing datum **DTM1** which passes through the center of the disk and the second arc.

Select the *Datum Plane* button in the right toolbar. Select the curved surface of the cutter on the side of the second arc. It highlights in red. A preview datum will show up in yellow. The default is a **Through** constraint, going through the central axis of this surface. Hold down the CTRL key and click on **DTM1**. The preview datum is now normal to DTM1 and still through the center of the surface. In the **Datum Plane** dialog window, go to the curved surface reference and click on the *Through* constraint. In the pull-down list, select *Tangent*. The previewed datum now moves to be tangent to the cutter and normal to DTM1 - exactly what we want. See Figure 19.

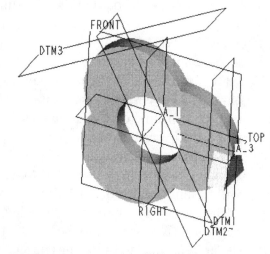

Figure 19 Tangent datum plane for second tooth

Accept the new datum by selecting **OK**. The new datum is called DTM3.

Now we'll create a one-sided solid protrusion on the new datum plane. Select the **Extrude** tool and **Activate Sketcher** in the dashboard. Pick DTM3 as our sketching plane. The yellow arrow shows the direction of view onto the sketch plane. Flip this so that it points away from the center of the cutter. The sketch orientation reference is DTM1 and it should face the Top of the sketch. Now select the **Sketch** button.

The cutter will re-orient. You might like to give the part a small spin to make sure you understand its orientation. You may also find it easier to sketch when the display is set to wire-frame or hidden line. We're going to create the sketch shown in Figure 20.

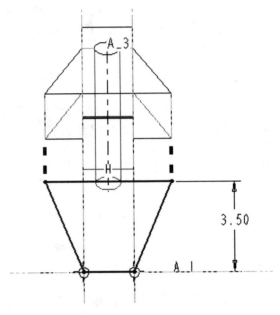

Pick the following five sketching references: DTM1, both sides of the disk, and the two outer edges of the first tooth. Now create the sketch shown in Figure 20. This sketch only needs one new dimension because the lines and vertices snap and/or align to the various references. When the sketch is complete, leave Sketcher and choose a **Blind** depth specification and enter the value **2**.

Verify that the tooth is the correct geometry and **Accept** the feature.

Figure 20 Sketch for second tooth

Create another **Straight Linear** hole using the outer planar surface of the tooth as the primary reference (placement plane). Use the datums **FRONT** and **DTM1** for the linear placement references (giving a dimension of *0* to each). The hole has a diameter of *1*. Specify a **Blind** depth and enter a value of *8.0*. Once again, if any of the dimensions are hard to pick on the screen, you can set these in the dashboard or the slide-up panels. The completed tooth looks like Figure 21.

IMPORTANT NOTE:
    Although this results in exactly the same geometry as the first tooth, notice our change in design intent. This tooth is to go a specific depth into the disk measured inwards from the circumference rather than outwards from the center. In this way, the tooth will be tangential to the disk regardless of the disk's size. Similarly, the hole's depth is a fixed value into the disk. At the present time, the hole goes through the surface of the inner hole. We will examine the effects of this later.

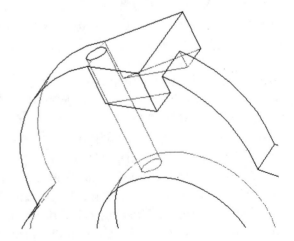

Figure 21 Second tooth completed

Third Tooth - Using Make Datums

The model is getting pretty cluttered up with datum planes which is making it more difficult to pick things out on the screen. One way to deal with this, of course, is to just turn off their display. This gets rid of them all, which cleans up the display but may make selecting them more difficult (you could always use *Search* to find them or use the model tree). A more selective way of controlling their display is to *Hide* them. Do that now with datum DTM1 - select the datum and then at the bottom of the RMB pop-up menu select *Hide*. The datum disappears (but not its children!). Do the same with DTM2 and DTM3 (use CTRL-click to select them both at the same time). The default datums should still be visible. Open the model tree and observe the gray box on the icons for these datums that indicates they are hidden.

All the datums we have created up to now have taken their expected place on the model tree. If a datum is only going to be used once to create another feature, it seems wasteful to create one that will be a stand-alone permanent feature on the model tree. Furthermore, we would likely want to Hide it to get it out of our way. The solution used by Pro/E for both these problems is a *make datum*. This is a datum that is created just when needed ("on-the-fly"), and then is automatically hidden when the feature using it is accepted. Make datums are sometimes called "datums-on-the-fly" for precisely this reason. The new terminology for make datums in the Wildfire release is "asynchronous datums" which is a bit of a mouthful. We will continue to use the old terminology. One other new facet of make datums is that they are listed on the model tree, but in a special way which we will soon discover.

The rules and methods for constraining a make datum are the same as if it was a regular one. What determines whether a plane is considered a make datum is *when* it is made. All our previous datums were created *before* we launched the commands that used them as references. For example, for the first tooth we created DTM2 first, then picked Extrude, then identified DTM2 as the sketching plane. For a make datum, this sequence would be changed: pick the extrude command first, then when we are asked to identify a sketching or reference plane, make the datum "on-the-fly". This is sort of a "just-in-time" delivery notion.

We are going to do other things in a slightly different order here, by creating the hole first. However, a hole requires a planar surface for its placement plane. We don't have such a plane at the desired angle. So, we will create the hole using a make datum to act as the placement plane.

Proceed normally to start the hole creation - that is, select the *Hole* toolbar button. You are asked (see the message area) to select a placement plane - but there isn't one in a suitable orientation. Here is where we will make the datum on-the-fly. Select the *Datum Plane* toolbar button. The Hole dashboard is grayed out, and the **Datum Plane** dialog window appears. We need to specify the constraint references for the new datum. Select the reference A_1 of the cutter. This is entered in the **Datum Plane** window with the default constraint *Through* (just what we want). Now CTRL-click on the TOP datum. It is added in the **Datum Plane** window with the *Offset* constraint with some rotation angle assumed. Change this angle to 30 degrees below the TOP datum, as shown in Figure 22 (you may have to use a negative angle). When this feature is finished, select *OK*.

We can continue on with our hole creation. Select the *Resume* button at the right end of the dashboard (the only button active) to return to the hole creation. A previewed hole will appear on the new datum plane (which is called DTM4). Set the diameter to *1.0* and the **Depth Spec** to *Through All*. Now drag the linear reference drag handles to FRONT (you can drag to anywhere along the displayed edge) and A_1. The distance from each reference is *0.0*. You may find it easier to set these in the **Placement** slide-up panel. The hole preview is shown in Figure 23.

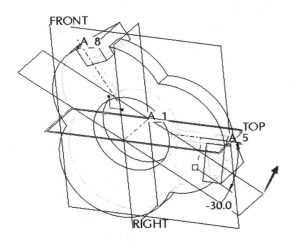

Figure 22 Creating the Make Datum

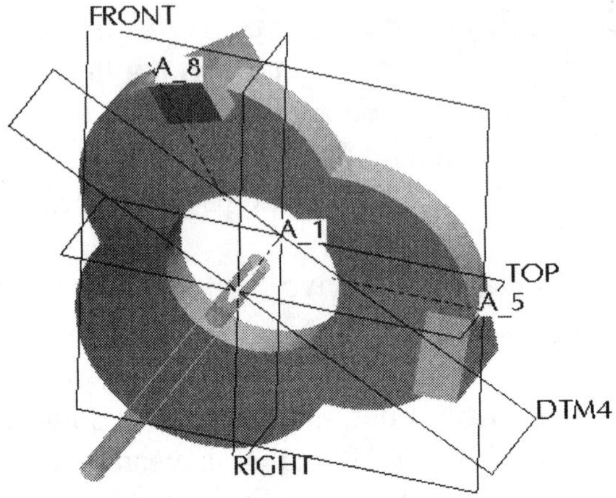

Figure 23 Creating the hole on DTM4

When the hole is accepted, there is no sign of the make datum we just created, although the hole does have an axis. Open the model tree. The other datums are all there (some are hidden), but the make datum DTM4 is not. However, the last feature on the model tree must be the hole we just created. This is shown as a group. Open the group by clicking on the "+" sign, and there is our hidden DTM4 along with the hole that used it. Right click on DTM4 in the model tree and select *Edit*. You will see the angle dimension associated with this make datum.

Now create the last tooth. This will also be a sketched protrusion on a make datum that is perpendicular to the axis of the hole (that's why we made it first) and tangent to the cutter surface. Select *Extrude > Activate Sketcher*.

The **Section** window is waiting for us to specify the sketching plane. Move it (the Section window) out of the way and select the *Datum Plane* button. Select the axis of the hole we just made, and (using CTRL-click) the surface of the cutter where the hole comes out. In the **Datum Plane** dialog window, set the constraints for these references in the pull-down lists to *Normal* (for the axis) and *Tangent* (for the surface). The preview should show a datum plane at the correct location and orientation. See Figure 24. Accept this datum and return to the **Section** window.

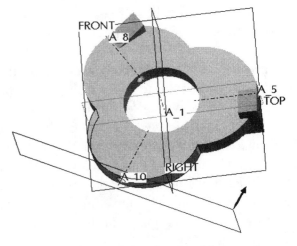

Figure 24 Make datum for the third tooth

Set the view direction away from the center of the cutter using *Flip*. Use **FRONT** as the *Left* orientation reference plane for the sketch. Once again, check your view orientation relative to the part. Pick the existing tooth edges as references, plus the front and back of the cutter body. Since these are all parallel, you are still only **Partially Placed**. For the final sketch reference, pick axis A_1. Sketch the tooth as shown in Figure 25. Note that, in order not to fill in half the hole through the tooth, we must sketch around the circumference of the hole. The *Use Edge* button

is handy for this. Select this command and pick on the curved edge of the hole. You may have to pick edges on each side of the circle, and come back later and trim edges away.

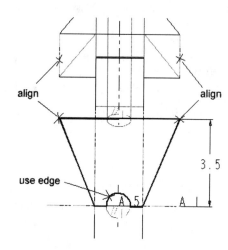

Figure 25 Sketch of third tooth

When you are finished with the sketch, select a *Blind* depth of *2*. We have now finished constructing the part, which should look like Figures 26 and 27.

Open the model tree and check how the last tooth is represented. In particular, where is the datum we used for the sketching plane and how is it shown? Save the part!

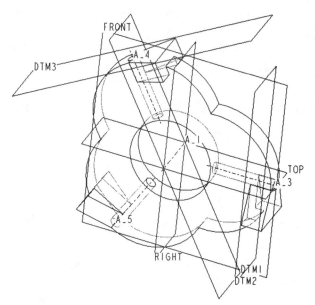

Figure 26 Finished part - wireframe showing datums

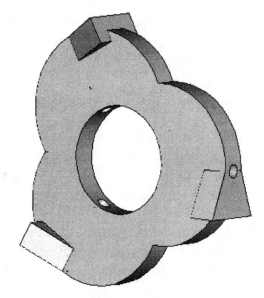

Figure 27 Completed part

Exploring the Model

We have created three geometrically identical teeth using three different modeling strategies. Let's see what happens when we start to play with the dimensions of the features. Try the following and see if you can explain what is going on. Before you try any of this, save the part so that you can recover from any future disasters! In each case, change the geometry back to the original before making a new modification.

1.   Change the radius dimension of the first circle of the disk (currently 6.0) to values of **5.0** and **8.0**. What happens to each of the tooth/hole features? Why?

2.   Change the diameter dimension of the construction circle (currently 8.0) to values of **4.0** and **12.0**. What happens? Why?

3.   Change the diameter of the large coaxial hole to 0.5. What happens? (This is easier to see in hidden line.) Why?

4.   What happens if you try to *delete* the datum **DTM1**? What about **DTM2**? Don't actually delete the datums.

5.   Turn all the datums off. Use the *Search* command to find **DTM1**. What happens if you try to change the angle of the datum **DTM1** to 45°?

6.   Here is an interesting genealogical problem: Can a feature be listed as its own parent? As its own child? Check the parent/child information window for the first cutter tooth and explain what you see there.

7.   Examine the parent/child relationships in the model. It is possible that, rather than being related only through the width alignment, some of the tooth/holes refer to other features in ways that were not intended. A possible reason for this is when you were selecting references, the picks were made to axes or edges of previously created features rather than the datum planes or surfaces. How you can be more selective in choosing references?

8.   Can you change the offset of **DTM2**? What happens if you specify an offset of *6.0* or *12.0*?

9.   Can you change the diameter of the hole going through the third tooth? Where does the dimension appear for this? What happens if you set this diameter to *1.5* and *2.0*? Why?

10. Can you change the angle of the Make Datum used to create the third hole? What happens if you change this angle to 60°?

11. Can you change the depth of the second and third teeth easily?

12. Suppress the central hole. What happens to the small radial holes? How far through does the first one go? Does anything happen to the third tooth? Why?

13. How many independent dimensions are there in this model? What is the minimum number that should be required? Set up the model so that only these dimensions can be modified.

14. Of the three methods used to create the teeth, which one would you say is the "best"? Keep in mind our three modeling objectives (simple, robust, flexible).

15. Can the *Search* command be used to locate Make Datums?

Considering *Design Intent*

You should be able to see once again that capturing the design intent is an important part of feature-based modeling and the model creation strategy. In this lesson we have added an important new consideration to this strategy - datum planes. Design intent involves consideration of the following:

- What is the design function of the feature?
- How does this influence the modeling strategy?
- How does the design function of a feature relate to other features?
- Which features should be unrelated in the part?
- How can you set up references and dimensioning schemes so that the parent/child relations reflect the above?
- How can you create the model so that it is driven by as few as possible dimensions? Will this necessarily always be desirable?
- When should you use relations internally in the part to drive the geometry automatically, depending on the critical design dimensions?

Design changes are inevitable. Therefore, you should try to design the model so that it will be easy to make the kinds of changes you expect in as direct a manner as possible. This is hard to do if you know only a few methods to create new features since your choices will be limited. You can often create the correct geometry, but it may be very difficult to modify or change later. Furthermore, it is often difficult to foresee exactly how you might want the model to change later. One thing is for sure, if you just slap-dash your features together, sooner or later you will run into a serious modeling problem that can become a nightmare for making design changes.

In the next lesson we will look at commands for creating patterned features (linear and radial patterns) and several ways of making feature copies. There will be more discussion of feature groups. We'll also see some new Sketcher tricks and a new type of protrusion (*Thin*).

Questions for Review

Several of these questions will require you to do some exploring of the program on your own.

1. What are the constraint types for creating datum planes? How are these different for make datums?
2. What are the constraint types for creating datum axes?
3. What combinations of constraints will lead to a completely constrained datum plane? Draw some freehand sketches to illustrate these.
4. What references are required to create a coaxial hole?
5. For a **Symmetric** solid protrusion, do you specify the depth in each direction, or the total depth? What about for a **Symmetric** cut?
6. What is the easiest way to create a datum plane parallel to a previous one at a specified distance away? What is this called?
7. Suppose you want to create a datum plane at an angle to another datum. You want to use the **Through** placement option, but there is currently no part edge or axis to use as a reference. How can you create the desired datum?
8. Does the order of selection of **Through** and **Offset(Rotation)** matter?
9. What is the difference between **Through All**, **To Next**, and **Up to Surf** when specifying an extruded feature's depth? For the last two, what happens if the extruded sketch does not completely intersect the specified surface?
10. Compare the advantages and disadvantages of using permanent datums and make datums.
11. If you want to *Edit* a feature created using a make datum as a sketching plane, where does the sketch show up?
12. Can you use the *Edit Definition* function on a hidden feature (like a make datum)?
13. When the model starts to get cluttered up with surfaces, edges, datums, and axes, how can you make sure that you are making an alignment to the desired entity?
14. Can the dimensions of a make datum (offset distances or angles) be controlled using relations?
15. What is the difference between the symbols "**dx**" and "**sdx**"?
16. Are other feature dimensions available for use in Sketcher relations?
17. Find a simple mechanical part and try to "reverse engineer" the design intent. How would you implement this in Pro/E?
18. Where and how do make datums show up in the model tree?
19. Can you use a make datum as a reference for another make datum? That is, can make datums be *nested*?
20. Is there an axis equivalent to a make datum (sort of a "make axis") that behaves in the same way as a make datum? That is: you make it on-the-fly, it is automatically included in the feature group, and its display is automatically hidden.
21. Does the *Dynamic Trim* tool affect construction lines?

Exercises

Here are some objects for you to make. Don't worry about exact dimensions, but datums and make datums will come in handy for these!

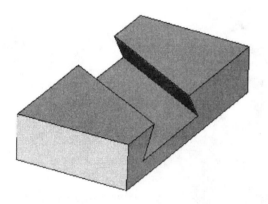

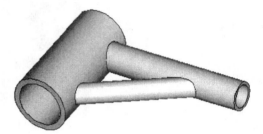

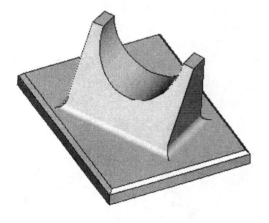

Project

Here is another part for the project. More pictures with dimensions are shown on the next page (all dimensions in millimeters). As usual, study the geometry carefully, and plan your modeling strategy before starting to create anything!

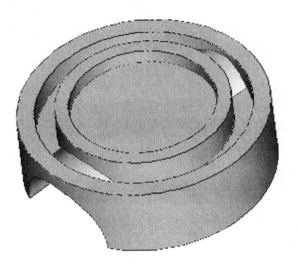

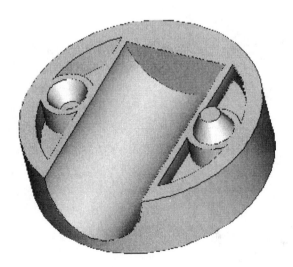

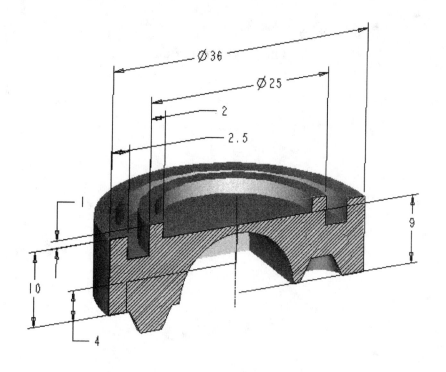

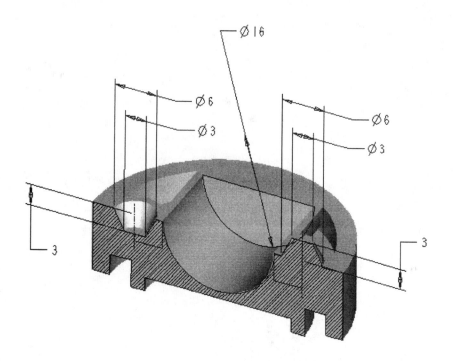

This page left blank.

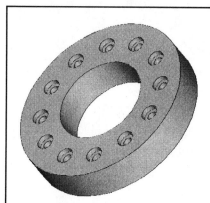

Lesson 7

Patterns and Copies

Synopsis

Naming dimension symbols. Uni- and bi-directional patterns. Creating a counterbored hole; hole notes. Radial patterns of placed and sketched features. Group patterns. Copies using translation, rotation, or mirroring.

Overview of this Lesson

Models often contain repetitive instances or copies of the same geometric form. In Pro/E terms, these are called *patterns*. As the name suggests, a pattern represents a regular, geometrically repeated placement of features. We will find that patterns can do a lot more than this. There are numerous options available when creating patterns. In this lesson we will look at the simpler and more commonly used pattern types.

When only a single duplicated feature is desired, it may make more sense to copy it rather than create a pattern. The copy command also allows more freedom in selecting references for the duplicated feature, including making it independent of the original. Copies can be made by translation or rotation (or a combination of these) or mirroring.

Patterns and copies work on individual features and on features arranged in groups. We'll look into the formation of groups a bit later in the lesson. To demonstrate the use of patterns and copies, we will be creating several different parts. The parts are totally independent of each other, so you can jump ahead to any one of these:

1. Patterned Features
    ▸   simple uni-directional patterns
    ▸   bi-directional patterns
    ▸   radial pattern of holes with relations
    ▸   patterns of grouped features
    ▸   radial patterns of sketched features
2. Copied Features
    ▸   *Same Ref* copy

- ▸   *Translate* copy
- ▸   *Rotate* copy
- ▸   *Mirror* copy
3.   Design Considerations
    - ▸   some things to think about when designing with complex features

The use of named dimension symbols is helpful when dealing with patterns (and elsewhere in Pro/E). We will see how to do that first. As usual, there will be some Questions for Review, Exercises, and a Project part at the end.

Patterned Features

Patterns are created by making duplicates of an existing feature - called the *pattern leader*. There are four kinds of patterns in Pro/E:

- ▸   dimensional patterns
- ▸   fill patterns
- ▸   table-driven patterns
- ▸   reference patterns

In this lesson we will look at dimensional patterns. There are enough variations of these to keep us busy for a while. The simplest dimensional pattern is created in one direction by incrementing a dimension that locates the pattern leader on the part. Each increment of the pattern dimension produces a new *instance* of the feature at the incremented location. Patterns are even more powerful than just creating multiple instances: it is possible to form the pattern in two directions simultaneously and to change the geometry parametrically of each instance in the pattern. While the dimension that locates the feature is incremented, other dimensions of the pattern leader can be incremented so that the instances change size and/or shape. It is even possible to change size and shape of an instance without changing its location (see the exercises!). All instances in the pattern can be modified simultaneously, if set up to do so.

In the examples below, we will explore basic pattern techniques. There are many advanced uses of patterns which are not covered here[1].

Naming Dimension Symbols

Create a new solid part called ***pattern1*** using the default template. ***Delete*** (or ***Hide***) the default coordinate system. Create a base feature using an extruded protrusion. The Sketch plane is the TOP datum. As shown in Figure 1, the part is a 12 X 20 X 2 rectangular solid.

---

[1] See the on-line help for pattern tables, fill patterns, reference patterns, pattern relations, *Identical*, *Varying*, and *General* patterns. These are also discussed in the *Pro/E Advanced Tutorial* from SDC.

Now create an extruded protrusion near the front left corner of the base feature. This is a cylindrical protrusion of diameter 2 and height 1. See Figure 2 for the dimensions. This protrusion will be our pattern leader for the next several exercises.

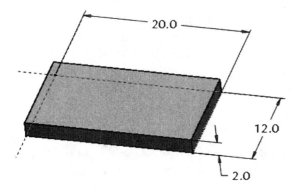

Figure 1 Base feature for part *pattern1*

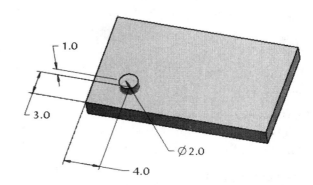

Figure 2 Dimensions of the pattern leader

In preparation for the pattern exercises, it will be helpful to change the symbolic names for the dimensions shown in Figure 2. To see the current names, double click on the protrusion and select

Info > Switch Dimensions

The symbolic names "*dxx*" will be displayed. Let's make those symbols a bit more meaningful. Preselect the horizontal dimension (4.0 in Figure 2). When it highlights in red, use the RMB pop-up and select *Properties*. Scan over the contents of the **Dimension Properties** dialog window that opens. At the top of this window, select the tab **Dimension Text**. There are three text entry areas at the bottom of this window. The name of the dimension symbol is shown. Enter a new name here - "*X*". Select *OK*. Notice on the part that the "*dxx*" symbol has changed to "*X*".

Change the depth dimension (3.0 in Figure 2) text to "*Y*".

Change the protrusion diameter (2.0 in Figure 2) text to "*DIA*".

Finally, change the protrusion height dimension (1.0 in Figure 2) to "*H*".

The protrusion dimensions should now look like Figure 3.

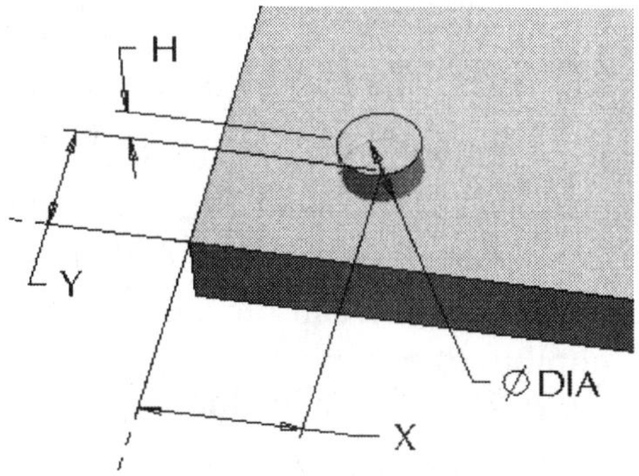

Figure 3 Renamed dimension symbols

You can use any alphanumeric symbol for a dimension. Keep the name short and don't use spaces or punctuation.
If a renamed symbol is used in a relation, Pro/E will automatically update the relation to use the new symbol.

Creating a Uni-directional Pattern

Turn off the datum planes and preselect the cylindrical protrusion. With this feature selected, one of the right toolbar buttons has become live. This is the Pattern tool 🔲 . You can launch the command with this button, or by using the RMB pop-up command in either the graphics window or the model tree. Choose one of these options to launch the **Pattern** command.

You will now see the dimensions associated with the protrusion[2] and the **Pattern** dashboard opens, with the Pattern tool symbol in the top row. The pull-down list at the left end of the lower row of the dashboard contains the four types of patterns. Leave this set to **Dimension**. In the rest of this row of the dashboard, the "1" and "2" refer to the first and second directions for the pattern. In the dashboard, move your cursor slowly across the four text areas and read the pop-up messages. Two of these text areas are for entering the number of instances in each pattern direction. The other text areas tell you how many dimensions are being incremented in each direction. The default contents are: Direction 1 (2 instances, No Items), Direction 2 (2 instances, No Items). The item box for Direction 1 is selected (yellow box).

Pattern #1 - Click on the horizontal dimension (4.0) for the protrusion. The tooltip indicates this is dimension "*X*". A text box appears which allows you to enter the increment to be used with this dimension. Enter a value of *6.0*. This means the next instance will be 6.0 units over to the right from the pattern leader (that is, the 4.0 dimension is incremented by 6.0). The one after that will be another 6.0 units over, and so on. In the dashboard, find the box for the number of instances to be created in the first

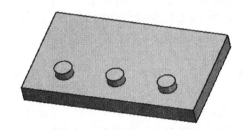

Figure 4 Simple pattern #1

pattern direction. Enter *3* in this box. We have now provided enough information (dimension to be incremented, increment size, number of instances) to create our first simple pattern. There is no preview or verify function with patterns. So cross your fingers and *Accept* the pattern definition (shortcut: middle click). The part should look like Figure 4. All three protrusions are highlighted in red as the last feature(s) created.

If you double-click on the second or third protrusion in this pattern, you will see all the dimensions associated with that instance. In particular, you can see the increment 6.0, and the number of instances. Change the increment from 6.0 to *4.0*, and the number of protrusions from 3 to *4*. Now *Regenerate*. Change the increment back to *6.0* and *Regenerate*. For this feature, you can even run the pattern off the right end of the part. This may not always be possible, especially if the feature being patterned was created with an open sketch. Running the pattern off the part can be a subtle

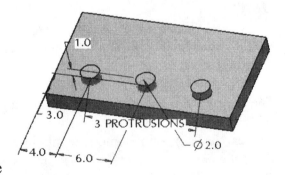

Figure 5 Pattern #1 dimensions

---

[2] If the dimensions are still in symbolic form, select *Info > Switch Dimensions*.

error to catch if you are creating a pattern of cuts or holes.

Open the model tree and expand the pattern. You will see each feature instance listed.

Pattern #2 - Let's play with this simple pattern some more. In the model tree, select the pattern and in the RMB pop-up select *Edit Definition*. This re-opens the pattern dashboard and shows all the dimensions for the pattern leader.

Open the **Dimensions** slide-up panel. Our dimension "*X*" is indicated in direction 1. Select this and use the RMB pop-up, select *Remove*. In its place, click on the dimension 3.0 for the protrusion - this is the dimension "*Y*". Enter an increment of *6.0*. In the dashboard, specify *2* instances. *Accept* the pattern. See Figure 6.

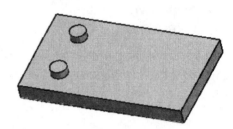

Figure 6 Simple pattern #2

So, the Direction 1 does not always have to be in the same direction on the part. In fact, the 1st and 2nd pattern directions are not physical directions at all, but more logical ones. The directions are a way of organizing which dimensions will be incremented together. It may even be that the feature being patterned does not move at all, but just changes shape and size (see the pyramid exercise at the end of the lesson).

Can a pattern go only in the "*X*" and "*Y*" directions? Let's find out.

Pattern #3 - Highlight the pattern in the model tree and using the RMB pop-up, select *Edit Definition*. Open the **Dimensions** slide-up panel. The "*Y*" dimension is listed in Direction 1. Click in the panel for Direction 1 and then CTRL-click on the horizontal dimension to add it. Set the increment to *6.0*. The panel should look like Figure 7. *Accept* the pattern (Figure 8). So, clearly, we are not restricted to incrementing a single dimension in pattern "direction". For that matter, we can increment in more than one physical direction at once. Let's explore this a bit more.

Direction 1	
Dimension	**Increment**
y:F7(PROTRU...	6.0
x:F7(PROTRU...	6.0

Figure 7 Pattern dimensions for pattern #3

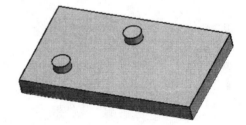

Figure 8 Simple pattern #3

Pattern #4 - Once again, select the pattern in the model tree and *Edit Definition*. Open the **Dimensions** slide-up panel. *Remove* the "*Y*" dimension using RMB. Holding down the CTRL key, click on the protrusion diameter dimension in the graphics window. Set an increment of *1.0*. Hold down the CTRL key and click on the height dimension. Set the height increment to *3.0*. Change the number of instances to *3*. The panel should now look like Figure 9. Accept the pattern and the model should look like Figure 10. The instances are created as before, but this time their diameter and height also change. Furthermore, it doesn't matter what order you specify the pattern dimensions shown in Figure 9. We could have picked the diameter dimension first, for example. Clearly, you can do more with patterns than just make duplicates!

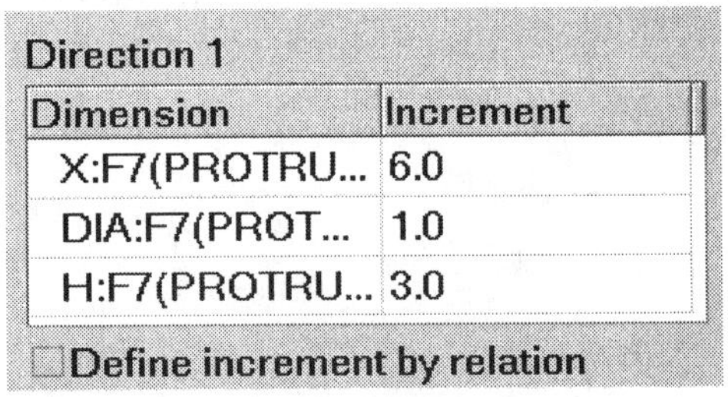

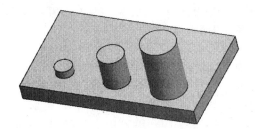

Direction 1

Dimension	Increment
X:F7(PROTRU...	6.0
DIA:F7(PROT...	1.0
H:F7(PROTRU...	3.0

☐ Define increment by relation

Figure 9 Pattern dimensions for pattern #4

Figure 10 Simple pattern #4

Creating a Bi-directional Pattern

So far, we have not done anything with the second pattern direction. In the model tree, in the RMB pop-up, select **Delete Pattern**. This removes the pattern instances but leaves the leader. (Be careful not to pick **Delete**, which gets rid of everything).

Pattern #5 - Highlight the cylindrical protrusion. In the RMB menu, select **Pattern**. Click on the horizontal dimension (4.0) and enter an increment of **6.0**. Down in the dashboard enter the number of instances as **3**. Now, still in the dashboard, click in the area for the second direction (farthest right box) that currently says "No items." When you click on the area, it turns yellow which means it is active. Now select the depth dimension (3.0) for the protrusion. Enter an increment of **6.0** and specify **2** instances.

Open the **Dimensions** slide-up panel (Figure 11). It shows "**X**" for Direction 1 and "**Y**" for Direction 2, along with their increments. **Accept** the feature (Figure 12).

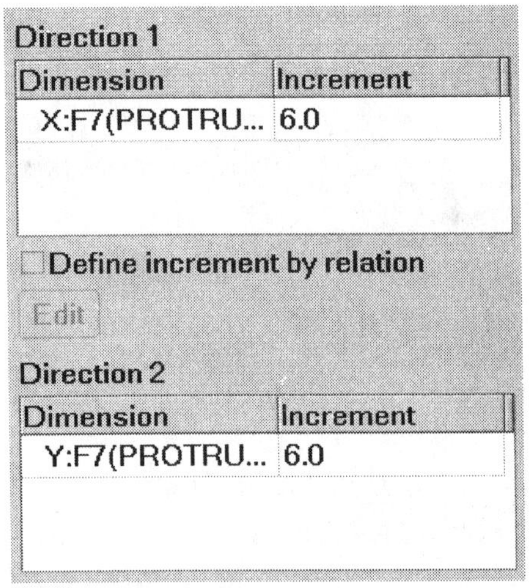

Direction 1

Dimension	Increment
X:F7(PROTRU...	6.0

☐ Define increment by relation

Edit

Direction 2

Dimension	Increment
Y:F7(PROTRU...	6.0

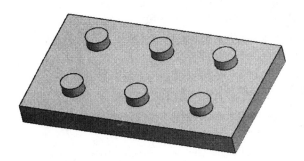

Figure 12 Simple pattern #5

Figure 11 Pattern dimensions for pattern #5

Let's modify the pattern one more time.

Pattern #6 - Select the pattern in the model tree and ***Edit Definition***. Click on the Direction 1 box that says "1 Item". It turns yellow (active). Hold down the CTRL key and add the diameter dimension (increment *1.0*) and the height dimension (increment *2.0*). Now click on the Direction 2 box that says "1 Item". Using the CTRL key, add the height dimension (increment *4.0*). Open the **Dimensions** slide-up panel to see all the pattern dimensions. See Figure 13. ***Accept*** the pattern, which should look like Figure 14.

Direction 1

Dimension	Increment
X:F7(PROTRU...	6.0
DIA:F7(PROT...	1.0
H:F7(PROTRU...	2.0

☐ Define increment by relation

[Edit]

Direction 2

Dimension	Increment
Y:F7(PROTRU...	6.0
H:F7(PROTRU...	4.0

Figure 13 Pattern dimensions for pattern #6

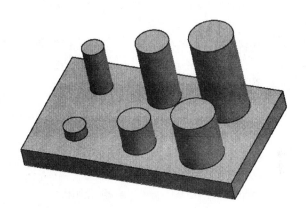

Figure 14 Pattern #6

We'll leave this part now. ***Save*** it and perhaps come back to it later to explore more simple pattern options. After you have saved it, select ***File > Erase > Current***. This removes the model from memory (takes it "out of session").

Creating a Simple Radial Pattern

A common element in piping systems and pressure vessels is a bolted flange. Here is how to create a flexible pattern of bolt holes. To demonstrate this, we'll explore the ***Hole*** dashboard to create a standard counterbored hole. In addition, we will set up a couple of relations to control the geometry based on the specified number of holes.

Start a new part called *flange* using the default template. Create the circular disk with central hole shown in Figure 15. We will need a central axis for the

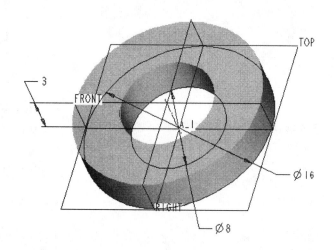

Figure 15 Base feature of flange

counterbored hole placement so you have a number of options: a) create a solid protrusion of two concentric circles, b) create a solid circular disk and add a coaxial hole, or c) revolve a rectangle around a central axis aligned with the datums. Each of these options will create the axis automatically. Pick whichever of these options you are *least* familiar with - might as well practice! The outer diameter is *16*, the hole diameter is *8*, and the disk is *3* thick. Note that the disk is constructed on **TOP**.

Now we'll create a single counterbored hole in the disk. This will be the pattern leader. In order to specify the pattern using an angular dimension, we choose a *Diameter* placement scheme (requiring an angle from a reference plane, and a diameter dimension for the flange bolt circle).

Select the *Hole* tool in the right toolbar. Moving across the dashboard from left to right, do the following:
- pick the **Standard Hole** button
- select a **UNC** thread type
- set size **1-8** (it's way down at the bottom of the hole list)
- set the depth to **Through All**
- leave the **Tap** button pressed
- turn off *Countersink*, and finally
- press *Counterbore*.

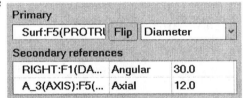

Figure 16 Placement panel for counterbored hole

Read the message window. Pick on the surface of the top of the protrusion at about the 5 o'clock position. A hole will be previewed. Open the *Placement* slide-up panel. Change the Primary placement type from *Linear* to *Diameter*. See Figure 16. Click in the secondary references pane. Drag one placement handle to the axis of the protrusion - a diameter dimension appears; drag the other to the RIGHT datum - an angle dimension will appear. Change the diameter dimension to *12* and the angle dimension to *30* as shown in Figure 17.

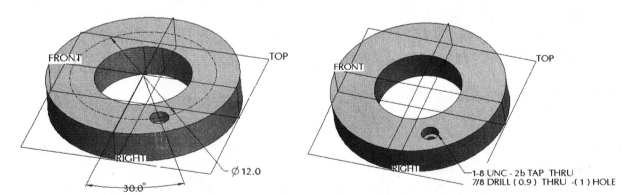

Figure 17 Placement dimensions for counterbored hole

Figure 18 Completed hole with note.

Open the **Shape** slide-up panel. Note the depth of the counterbore. Change the thread depth from *Variable* to *Thru Thread*. Leave the checked box beside *Include Thread Surface*.

Open the **Note** slide-up panel. This shows the text of a note that will be attached to the hole in the database. This can eventually be displayed on the part drawing.

Open the **Properties** panel. This contains a table showing all the parameters associated with this hole.

Accept the feature. The hole appears (Figure 18) with the hole note, as might appear in a drawing. Open the model tree and expand the entry for the hole. The note is listed there[3]. To move the note, select it in the model tree, use the RMB and select *Move*. You can click anywhere on the screen to drop the note. See what happens when you spin the model. To turn off the display of the note, select *Tools > Environment* and deselect the option **3D Notes**.

You can create 3D notes for other features as well. These notes can contain any text and are useful ways to attach documentation to the model. In the model tree, select the base feature, and in the RMB pop-up menu, select *Setup Note > Feature*. This opens the **Note** window. Basically, you need to enter the note text, then select the *Place* button to tell Pro/E how and where to attach the note. Come back later and try this (you will have to turn 3D Notes display back on).

Now, back to our pattern of holes. This first hole becomes the pattern leader. We are going to make a pattern of 8 instances of the hole spaced equally around the flange. That is, to create each instance, we will increment the angular placement of each hole by 45°. This is another example of the importance of planning ahead: if you are going to use a pattern, you have to have a dimension to increment! For example, we could not create the bolt circle if we had used a linear placement for the pattern leader (or at least it would be very difficult) or especially not if we had aligned it to either FRONT or RIGHT. As we create the pattern, follow the prompts in the message window. Select the hole and in the RMB pop-up, select *Pattern*.

This is actually a uni-directional pattern in disguise. There is only one dimension to increment - the angle to the pattern leader. Click on the 30 dimension, and enter an increment of **45**. Change the number of instances to **8**. *Accept* the pattern. See Figure 19.

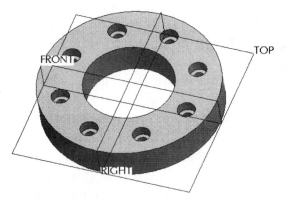

Open the model tree to see how the pattern is listed there. Does every hole inherit the 3D Note of the pattern leader?

Figure 19 Pattern of holes

Setting up Pattern Relations

This bolt pattern is not symmetric about the vertical datums. Also, suppose we wanted to change the number of bolts on the flange - this would change all the angular dimensions. Do we have to recreate this pattern from scratch? The answer is no - we can use relations! In the pull-down menu, select

> *Tools > Relations*

---

[3] You must turn on the display of notes in the model tree using *Settings > Tree Filters*.

Select the ***Insert Symbol from Screen*** button in the **Relations** toolbar (or use ***Insert > From Screen***) and click on the 2nd hole in the bolt pattern (the one at about 3-o'clock). You should see all the dimensions that control the pattern as in Figure 20. Note that some hole dimensions have been removed from this figure for clarity.

Take note of the symbols for the following dimensions (your symbols might be different): angular dimension between holes ($d11$), the angle of the first hole from RIGHT ($d5$), and the number of holes ($p12$).

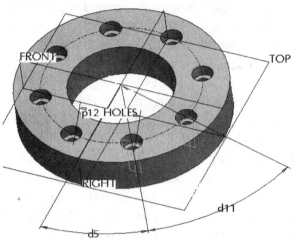

With these symbols identified, enter a couple of relations as follows (using your own symbols, of course):

> **/* angular separation of holes**
> **d11 = 360 / p12**
> **/* location of first hole from RIGHT**
> **d5 = d11 / 2**

Figure 20 Critical pattern dimensions for setting up some relations

(Remember that you can easily pick off the dimensions on the graphics window using the ***Insert From Screen*** button in the **Relations** toolbar.)

Note that the second relation uses a value computed by the first relation. In the database, all relations are evaluated top-down. Before you leave the **Relations** menu, select ***Switch Dims*** to return dimensions to numeric values. Also, select the ***Verify*** button. Select ***OK***. You must select ***Regenerate*** for the relations to take effect. The only dimension that will change this time is the angle to the pattern leader. This pulls the entire pattern around slightly. Double click on any of the holes and change the number of holes to *12*, then ***Regenerate*** the part. See Figure 21. Check again for 6 holes, 5 holes. Don't forget you have to regenerate after each modify. In each case, the correct number, separation, and pattern leader placement are automatically determined.

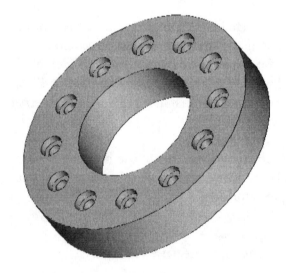

Figure 21 Hole pattern modified for 12 bolts

Double click on any of the holes, and change the diameter of the counterbore to *2*. ***Regenerate*** the model. Note that all pattern members change. With the diameter of 2, try to create a pattern of *16* holes, then *20* holes. What happens?

If you have been in shaded mode, go to ***Hidden Line***. What do you see on each hole, and what does this mean? Select the pattern leader and ***Edit Definition***. Open the **Shape** slide-up panel and deselect ***Include Thread Surface***. ***Accept*** the feature. Note that Pro/E does not show the physical thread of a hole (that would take too much CPU resource!), but uses a visible indicator to show that a thread is defined. This is called a *cosmetic* thread. If you want to show the physical thread, you can cut one with a *helical sweep* feature.

If you want to play with this part later, then ***Save*** it now. Otherwise, select ***File > Erase***.

A Pattern of Grouped Features

The patterns we have seen up to now have involved a single feature as pattern leader. Sometimes, a geometric shape that you want to pattern requires several features to create. In order to pattern these, they must first be grouped together.

We are going to create the part shown in Figure 22. The pattern leader is on the left in the front row. Each instance in the pattern actually consists of three features: a protrusion, a hole, and a round. We will use a pattern to set up two rows with the dimensions of the features incrementing along each row, and between rows. This is the same as the bi-directional pattern we did before, but this time involving three features simultaneously.

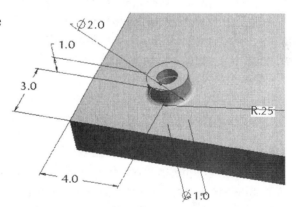

Figure 22 Pattern of grouped features

Open the part ***pattern1*** that we used before. Delete the previous pattern, keeping the pattern leader (the cylindrical protrusion). Create a ***Through All***, coaxial hole (diameter *1.0*) on the axis of the protrusion, and a round (radius *0.25*) on the edge around its base. The dimensions are shown in Figure 23.

Before we can create the pattern, we have to group all the features (circular protrusion + hole + round) into a single entity - called (no surprise!) a *group*. Note that grouped features must be adjacent to each other in the model tree. In the pull-down menu, select

Figure 23 Pattern leader composed of circular protrusion, hole, and round

Edit > Feature Operations > Group

Ignore the window that opens and continue picking commands in the cascading menus at the right.

Create > Local Group > [holder]

where *holder* is the name we supply for the group in the message window. Now pick on the hole, the protrusion, and the round (remember to hold down the CTRL key while you select these). You may want to zoom in on the model to make sure you select the right features. Or, you can select the features in the model tree; each will highlight when selected. Then select *OK > Done*

You should be informed that the group *holder* has been created. In the model tree, the features are now included inside a group of the same name. Now, still in the **GROUP** menu at the right of the screen, select *Pattern* and pick on the protrusion. The pattern dashboard opens and you should see all the dimensions associated with the group as shown in Figure 23. We now can select the various dimensions we want to increment in the first and second directions. Select the following (hold down the CTRL key while selecting dimensions!):

First Pattern Direction
1. pick on the 4.0 dimension, and enter the increment *6.0*. This will increment the location of the group along the plate.
2. pick on the diameter of the protrusion 2.0, and enter the increment *1.0*
3. pick on the height of the protrusion 1.0, and enter the increment *1.0*
4. pick on the diameter of the hole 1.0, and enter the increment *1.0*
5. enter the number of instances *3*

Now left click on the item collector for the second direction.

Second Pattern Direction
1. pick on the 3.0 dimension, and enter the increment *6.0*. This will increment the location of the group to the next row.
2. pick on the height of the protrusion 1.0, and enter the increment *4.0*
3. pick on the protrusion diameter 2.0, and enter the increment *1.0*
4. pick on the hole diameter 1.0, and enter the increment *1.0*
5. enter the number of instances *2*

For a summary of all this, open the **Dimensions** slide-up panel. *Accept* the pattern.

What dimensions are available for modification (this may depend on which feature you pick)?

What happens here if you try to create a group off the end of the plate by extending the pattern? Change the direction 1 increment to *10* to find out. How do you recover from this?

Open up the model tree to see how a group pattern is represented.

Save this part and erase it from the session.

Radial Patterns of Sketched Features

A common modeling problem involves creating radial patterns of sketched features. The radial hole pattern we did earlier was of a placed feature, and was pretty easy. For sketched features, we must be a bit more sophisticated in order to create an angular dimension which can be incremented to produce the pattern. The most common way of doing this is by using make datums in the creation of the pattern leader[4]. We set up the feature so that the make datum is used as a reference for the sketch. The make datum is typically created using ***Through*** and ***Offset(rotation)*** options. Then, we can increment the rotation angle to form the pattern. It is crucial when creating the sketch of the feature that there are no references, alignments, or dimensions to entities (datums, edges, surfaces) in the part other than the make datum, the axis of the radial pattern, or other axisymmetric features.

There are two types of these radial patterns. In the first, the feature is sketched on a plane that goes through the radial pattern axis (or one parallel to this); the feature extrudes in a direction perpendicular to the axis. In the second, the feature is sketched on a plane normal to the axis and extrudes parallel to it. We will see examples of both types here. In the first, a make datum is used as the *sketching plane* for a protrusion. In the second, the make datum is used as the *sketching reference plane* for a cut.

Radial Pattern using Make Datum as Sketching Plane

The first part we will make is (very approximately) the geometry of a turbocharger rotor. We will take this opportunity to introduce a new feature variation - a *Thickened* extrusion[5], and a new arc type in Sketcher.

Start a new part called **turbo**. Use the solid part template for **millimeter-Newton-seconds**. Close the model tree.

Our base feature is a revolved solid protrusion. Sketch this on **FRONT**. The sketch is shown in Figure 24. The curved edge in this sketch is a conic arc (use the button on the end of the arc fly-out). Click at the two are end points and drag out the arc until each end shows a tangent constraint; then drop the arc. The conic dimension (0.4 in the figure) is not a radius but a parameter (rho) used to define the conic section. For a perfect elliptical curve, this parameter should be 0.414... (enter this value as [**SQRT(2)-1**] and note the use of the built-in math function). This expression is automatically inserted as a sketch relation. Notice that the constraints are

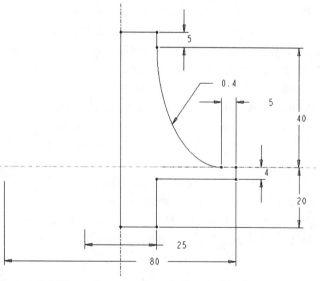

Figure 24 Base feature sketch for turbocharger

---

[4] We prefer to use a make datum because we don't want the screen cluttered up by a bunch of duplicated datum planes.

[5] This is another change in terminology introduced in Wildfire. Previously, this was known as a *Thin* solid.

turned off in the figure here for clarity. Don't forget the centerline (lined up with **RIGHT**). Create this sketch and then revolve it through 360° to get the shape shown in Figure 25.

Now we will make our first turbocharger blade to serve as pattern leader. We are going to create a sketching plane through the axis of the revolved base feature at an angle to **RIGHT**. We can then create the pattern instances by incrementing this angle. Since the sketching plane is through the axis, the extruded blade will be normal to the axis. We are also going to use a new type of feature that we haven't seen before - a *Thickened* feature (instead of *Solid*). All we have to sketch is a single line representing the cross section shape of the blade.

Figure 25 Base feature

Select the *Extrude > Activate Sketcher*. We want to make a datum-on-the-fly here, so move the Section window out of the way and pick the **Datum Plane** tool. Pick on the axis of the revolve. This appears as a **Through** constraint. Holding down CTRL, select the **RIGHT** datum. This gives us the correct constraint. Change the **Offset** rotation value to **60** degrees. Accept this with **OK**.

Back in the **Section** window, select the **TOP** datum with orientation **Top**. Then select **Sketch**. We are now in Sketcher looking directly at the make datum. For our sketching references, select TOP, the revolve axis, and the top surface of the base feature (we can use this surface because it will be constant for all instances in the pattern). Create the sketch shown in Figure 26. This consists of a single vertical line and a circular arc[6]. When the sketch is complete, select *Accept* in Sketcher.

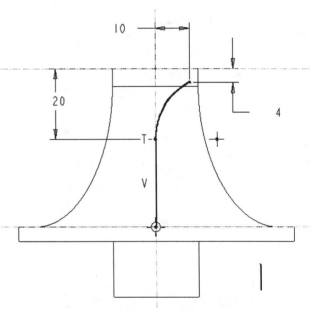

Since this was an open sketch, the preview consists of a single surface. A direction arrow points away from this surface that indicates the side Pro/E will add material. This obviously won't work here to create a solid.

Figure 26 Sketch for *Thin* protrusion - turbocharger blade

Pick the *Thicken* button on the dashboard, and specify a thickness of *1.0* (remember we are in mm). Beside this data field is a button that lets you specify which side of the sketched line to

---

[6] No claims are made here about the aerodynamic suitability of this blade shape, other than it's probably far from ideal!

add material (either side, or both). Watch the preview carefully as you cycle through these options. We want to thicken the sketch equally on both sides here. For the depth specification, select *Blind* with a value *60*. Verify the blade shape (see Figure 27) and *Accept* the feature.

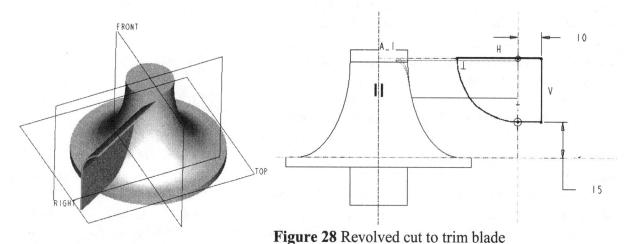

Figure 28 Revolved cut to trim blade

Figure 27 Completed blade

Before we complicate the part with the pattern of blades, we will create a revolved cut. This is sketched on another make datum that is *Through* the revolve axis and *Parallel* to one of the vertical flat surfaces of the blade. For the sketching orientation reference, select *Top > TOP*. In Sketcher, select as references the top edge of the blade, the outside edge, and axis (for the axis of revolution). Create the sketch shown in Figure 28. You might wonder why it needs to extend past the end of the blade. Come back later to change the dimension 10 to 0 and regenerate - don't do this now! Revolve this cut through 360° and accept the feature. Don't forget the *Remove Material* button!

We are now ready to pattern the blade. In the model tree, select the group containing the protrusion, and in the RMB pop-up menu, select

Pattern

Select the angle (60) used to create the make datum sketching plane. Enter an increment of *30* for this dimension. In the dashboard, set the number of instances to 12. That's all there is to this pattern. *Accept* the feature.

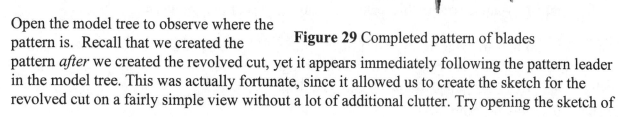

Figure 29 Completed pattern of blades

Open the model tree to observe where the pattern is. Recall that we created the pattern *after* we created the revolved cut, yet it appears immediately following the pattern leader in the model tree. This was actually fortunate, since it allowed us to create the sketch for the revolved cut on a fairly simple view without a lot of additional clutter. Try opening the sketch of the cut now to see what it looks like with all 12 blades in view.

Why could we not use a vertical surface of the blade as the sketching plane for the cut? Try that and see what problem arises. This will be more obvious if you make the blades thicker.

You might dress this part up with a coaxial hole and some rounds.

Let's look at another variation of radial patterns using a make datum. Save this part and then erase it from the session.

Radial Pattern using Make Datum as Reference Plane

In this example, the feature extrusion direction is parallel to the axis of the radial pattern. We must do something a bit different from the turbocharger. The main idea is the same - incorporate a make datum created using ***Through*** and ***Offset(rotation)*** into the sketch references. The angle parameter can then be incremented to produce the pattern. In the turbocharger, the make datum was the sketching plane; in this part, the make datum will be the sketching horizontal reference plane.

Start a new part **wheel_rim** using the default part template. Our base feature is again a revolved solid protrusion. The sketch plane is **TOP** and the axis of the revolve goes through **RIGHT**. The sketch is shown in Figure 30. Revolve the sketch through 360° and accept the feature.

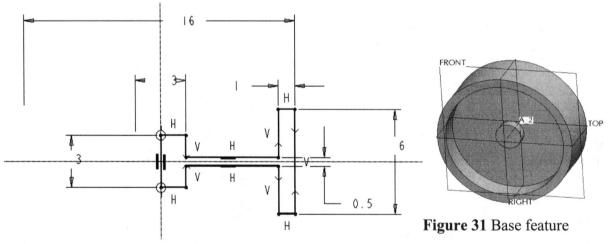

Figure 30 Base feature of **wheel_rim** - revolved protrusion

Figure 31 Base feature

We will create a pattern of cuts through the web of the wheel. The pattern leader will be a both-sides extruded cut, sketched on **FRONT** (the extrusion direction is therefore parallel to the axis of the wheel - compare this to the turbocharger where the extrusion direction was perpendicular to the axis). We want to pattern this feature around the axis of the base feature.

The most common error for this type of feature pattern is to use a *fixed* datum as the sketching reference plane, then a centerline in the sketch through the wheel axis and at some angle to this datum. If you then try to pattern this feature past 180° of rotation, the pattern will fail. This is because of the inner workings of Pro/E's geometry engine[7]. Try it sometime!

Select *Extrude > Activate Sketcher*. For the sketching plane pick **FRONT**. For the sketching reference plane, *Remove* the default RIGHT datum plane using the RMB pop-up. Instead, select the *Datum Plane* tool and create a make datum *Through* the axis of the wheel and *Offset* from the **TOP** datum. Enter an offset rotation angle of *30* degrees. The reference should face *Top*. Accept the make datum with *OK*, and you are in Sketcher.

We are now looking at **FRONT** with the make datum **DTM1** facing the top of the screen. The edge views of **RIGHT** and **TOP** are rotated a bit. We must be very careful now about picking sketch references. Pick only on **DTM1** and the axis of the wheel (when selected, you will see a small X there). Notice in the reference window that we are fully placed.

The sketch for the cut is shown in Figure 32. Be sure to avoid any dimensions or constraints that involve the fixed datums **TOP** and **RIGHT**. If you turn them off, you don't have to worry about this.

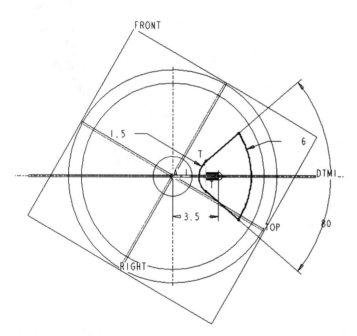

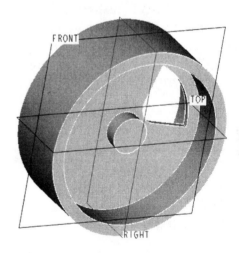

Figure 33 Pattern leader completed

Figure 32 Sketch of pattern leader (note rotation of TOP and RIGHT)

When the sketch is complete, back in the dashboard set the Depth Spec *Through All* on both sides. Select the *Remove Material* button and *Accept* the feature. See Figure 33.

We can now pattern the cut. Open up the model tree and right click on the group containing the cut. This group contains the make datum. In the RMB pop-up menu, select *Pattern*. In the dimensions on the screen, find the 30° dimension we used to locate the make datum reference

---

[7] There is a way around this problem using a construction line instead of a center line. Check it out.

plane from the TOP datum and click on it. Enter the increment *72°*, and specify *5* copies. There is nothing to do in the second direction, so just select *Accept*. The pattern should now be created as in Figure 34.

What happens if you try to make a pattern of 6 cuts at 60° increments? Would you say this is a *robust* model?

This concludes our discussion of patterns. There are many more things you can do with patterns, and some more advanced techniques. For example, instead of simply incrementing dimensions between instances, you can use pattern relations to develop formulas that will control the instance placement and geometry. Another tool called a pattern table allows you to place instances at non-uniformly spaced locations driven by dimension values stored in a table like a spreadsheet. A pattern fill is a new type of pattern introduced in Wildfire. This will duplicate features (like holes) in a regular geometric pattern in order to cover a region bounded by a datum curve. These advanced pattern functions are presented in the *Pro/ENGINEER Advanced Tutorial* from SDC.

Figure 34 Wheel_rim with radial pattern of cuts.

Copying Features

In the previous sections, we saw how to create a multiple-instance dimension-driven pattern of a single feature or a group of features. The pattern could only be created by incrementing one or more of the feature's existing dimensions. The **Copy** command allows more flexibility in terms of placement and geometric variation (you aren't restricted to the dimensions used to create a pattern leader, for example), but only creates one copy at a time. There are several options available with *Copy*, and we will create several different simple parts to illustrate these.

A *Same Ref* Copy

We are going to create the part shown in Figure 35. The vertical plate on the left is the original, and the one on the right will be the copy.

Start by creating a new part **featcopy** with the default template. Create a rectangular solid protrusion on **TOP** that is **10 x 20 x 2 thick**. Line up the left face of the block with the **RIGHT** datum, and the back face with **FRONT**.

Figure 35 Part with copied feature

For the first vertical plate, the sketching plane is a make datum that is *Offset* from **RIGHT** by *5*. Select the top of the rectangular base as the *Top* reference plane. Then sketch the protrusion as shown in Figure 36. Note the sketching references. The hole is included in the protrusion - Pro/E will know where to add material, and where to leave the hole. Also, the sketch must close across the bottom since you can't have a mix of open and closed curves in the same sketch (Try it!). The extrusion has a *Blind* depth of *1*. The part should look like Figure 37.

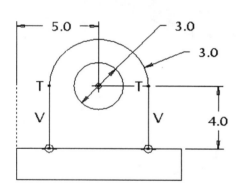

Figure 37 Sketch for vertical plate

Figure 36 First feature completed

Now, we are ready to copy the feature. We want the copy to be 10 units to the right of the first. If the geometry of the first feature changes, we want the copy to change too - it will be *dependent*. As you encounter these new menus, watch the message window and command description at the bottom of the screen. Launch the command sequence from the pull-down menu:

> *Edit > Feature Operations*
> *Copy > Same Refs | Select | Dependent | Done*

Pick on the vertical plate and select *OK > Done*.

The **GP VAR DIMS** menu will open up. This is giving us the opportunity to select which dimensions in the copy we want to vary from the original. At this time, we will only change the distance from the left end. Move the cursor up and down the listed dimensions. As you do this, the dimension will highlight on the model. Find the dimension *5* that locates the protrusion from the left end (this was the offset dimension for the make datum), and check it. Then select ***Done***. You are prompted for a new value for this dimension; enter *15* (note that this is a new dimension, not an increment as in a pattern), then in the **Group Elements** window at the right[8] select ***OK***. The new protrusion should appear at the right.

What happens if you try to ***Edit*** the hole diameter on the first protrusion? Or the height dimension on the copy? What happens if you suppress the original? The copy?

How does the copy appear in the model tree?

Delete the copy and create a new ***Same Refs***, ***Independent*** copy. Try the same modifications.

A Translated Copy

We will make the part shown at the right. The original feature is again in the lower left corner.

You can keep the same base plate as the previous part (10 x 20 x 2 thick, on TOP). You will have to delete the two vertical protrusions (or suppress them). Create a circular solid protrusion near the lower left corner of the plate (dimension *4* from left surface, *3* from front surface, diameter *3*, ***blind*** depth *5*). See Figure 39.

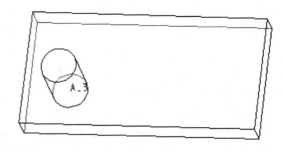

Figure 38 Part with a Translated copy

Now we will copy the feature and change its diameter at the same time:

> ***Edit > Feature Operations***
> ***Copy***
> ***Move | Select | Dependent | Done***

Click on the cylindrical protrusion, then ***OK > Done***.

> ***Translate > Plane***

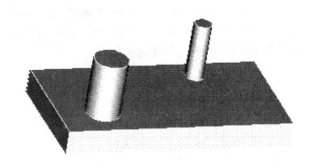

Figure 39 Original feature to translate/copy

---

[8] The *Elements* window is another hold-over from previous releases of Pro/E. Don't be surprised if it, too, disappears in the next few releases.

Select the RIGHT datum plane. Check the direction of the translation arrow. It should be pointing towards the right end. Select *Okay* and enter the distance *10*. The new feature won't show up just yet. To move it again:

> *Translate > Plane*

Pick on the front vertical surface of the plate. The default direction is the outward normal to a solid surface, so *Flip* the direction arrow and confirm the direction with *Okay*. Enter the distance *5*. Then select:

> *Done Move*

In the **GP VAR DIMS** menu, move the cursor down the list of dimensions and select the diameter of the protrusion as variable, then *Done*. Enter the new value *1.5* and *OK*.

Double-click on the copied cylinder. Spin the part and observe the location of the translation dimensions. The translation dimensions are displayed slightly differently - this is an easy way to pick them out. Note that the Move translation dimensions work somewhat like increments in a pattern. The main difference is you are not restricted to incremented dimensions of the pattern leader. You can move in any direction relative to any reference. Try moving the protrusion vertically.

You can now either *Save* this part or *Erase* it.

A Rotated Copy

Now, we will use a rotated copy to create the part shown in Figure 40 - a large circular pipe with two pipes joining it off-axis. At the same time, we will see a situation where feature creation order can be used to advantage (or foul you up!).

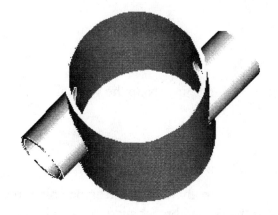

The original side pipe is on the left, the rotated copy is on the right. It can be obtained by a 180° rotation around the big pipe axis.

Create a new solid part **sidepipes** using the default template.

Figure 40 Part with Rotated copy

Start by creating a circular solid *both-sides* (symmetric) protrusion from the sketching plane **TOP**. Use **RIGHT** and **FRONT** as sketching references. Sketch a circle with a diameter of *20* and set *blind* depth of *20*.

Do not add the inner surface of the pipe at this time - we will do that later. This may not be an obvious thing to do but we have a situation where feature creation order is important as discussed below.

For the first side branch, create a one-sided solid protrusion. Use **FRONT** as the sketching plane (**Top** reference **TOP**) and sketch an *8* diameter circle *aligned* with TOP and with a center *5* from RIGHT (Figure 41). Check the feature creation direction arrow. Make the protrusion with a *blind* depth of *15.* This will extend it outside the circumference of the major pipe.

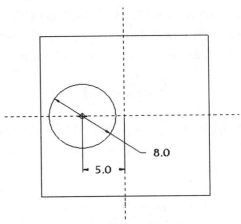

Figure 41 Sketch for side pipe

Create a *Straight, Coaxial hole* on the axis of the side pipe. The hole diameter is *7.* Use the placement plane **FRONT**. We want *Through All*. The part should look like Figure 42.

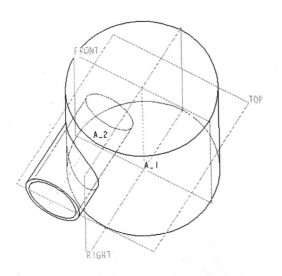

Figure 42 First pipe (Note the vertical pipe is still solid)

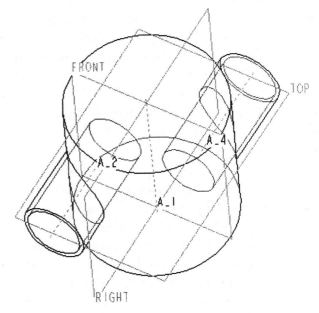

Figure 43 Rotated copy completed

Now we are ready to copy the branch pipe (protrusion plus hole). We may want to change the size of the copied branch pipe, so we will make an *independent* copy:

> *Edit > Feature Operations*
> *Copy > Move | Select | Independent | Done*

Pick on the side protrusion and hole (remember to use CTRL), then select *OK > Done*. Now we specify the rotation:

> *Rotate > Crv/Edg/Axis*

and pick on the axis of the main pipe. The red arrow shows the direction of rotation (right hand rule - it should be pointing up). Accept the direction with *Okay*. Enter the angle of rotation *180*. Then select:

> *Done Move > Done*

to keep all the existing dimensions. However, since we have created an independent copy, we could come back and change any dimensions of the copied pipe later. All the elements of the copy have been defined, so click *OK*. The model should now look like Figure 43.

Now we can add the central *Hole* of the main pipe. Make it a *Straight*, *Coaxial* hole from the placement plane **TOP**. Make it *Through All* in both directions, with a diameter of *19.*

This completes the model, so *Save* it now. We are going to explore it a bit, which might mess it up.

Exploring the Model

Now, you may be wondering why we left the central hole until last. Let's experiment with the *Edit* command (RMB pop-up menu), changing diameter dimensions of both the original and the copy. You can also modify the rotation angle. You should be able to modify both branch pipes with no problem. What happens if you modify the diameter of the main pipe to *12* and hole to *11?* The part will certainly regenerate, but is clearly wrong. However, the error is relatively easy to fix. Consider what would happen if we had used the following "obvious" sequence (what is important here is the order that features are created - you might like to sketch each feature in the following sequence as it is added to the part (or actually make a new part) in order to visualize the problem that would arise):

1.   **create main pipe** - same geometry as before.
2.   **create central hole** - same geometry as before (but in a different order).
3.   **create side branch** - We couldn't do this from **FRONT** since that would be inside the pipe (that now has the inner hole in it). We would have to create a **Make Datum** outside the pipe using an offset of 15 from **FRONT** and create the branch towards the main pipe using **To Next** or **To Selected** depth.
4.   **create the side branch hole** - We could use the planar face of the branch as the placement plane for a coaxial hole with a depth specified as **To Next** (through the next part surface encountered, ie the inside surface of the big pipe). Note that a *Thru All* would go out the other side.

These steps would create the same original geometry. However, we would have a big problem if we tried to reduce the diameter of the main pipe to anything less than 18, as we did above. Why? At step 3, the side branch solid protrusion would not totally intersect the surface of the main pipe as required by the **To Next/Selected** depth setting. The part would not regenerate at all, and we would have to spend some time fixing the model. This is a more serious problem than we have with the current model, which is therefore more robust. Once again, we see the need to plan ahead!

A Mirrored Copy

The final copy option we will look at is the mirror copy. We did this in an earlier lesson with a simpler geometry. Mirroring is very useful; obviously if you have symmetric parts, you only have to create half and then mirror to get the other half. We will create the simple mirrored, curved slot shown in Figure 44.

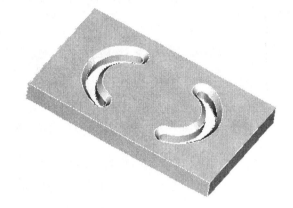

Figure 45 Part with Mirror copy

Create a solid rectangular base plate (part **mircopy**) of size 12 x 20 x 2, sketched on **TOP** so that the datum planes **RIGHT** and **FRONT** are on the centerline of the plate. Create a single *Through All* cut using the dimensions shown in Figure 45. Make sure in your sketch that all the arcs are tangential.

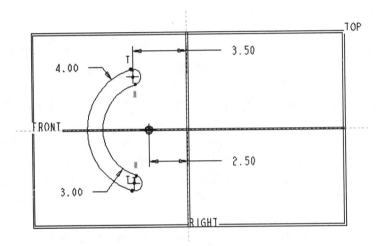

Figure 45 Sketch of original slot

Add a *45 x 0.5 Chamfer* to the upper and lower edges of the slot. Both edges can be in the same edge set. The mirror copy is easy:

> *Edit > Feature Operations*
> *Copy > Mirror | Select | Dependent | Done*

Pick on the cut and the chamfer (using CTRL), then select *Done*. The mirror plane is **RIGHT**. That's all there is to it! Try to *Edit* the mirrored copy. You should investigate to see what happens if you make an *independent* mirror copy of the same slot, and what happens if you try to make a mirror copy that intersects the original feature.

> **Helpful Hint**
>
> Here is a word of warning about *Mirror*. When you select this command, the *All Feat* option becomes available. Be aware that this grabs every current feature in the part and mirrors them. This includes all solid features, datum planes, datum axes, notes, etc....EVERYTHING! This is a "great" way to clutter up your model with useless and redundant features. This command has its uses, but must be used with discretion.

Design Considerations

We have covered a lot of ground in this lesson, and hopefully added a lot of ammunition to your modeling arsenal! We have also seen how the feature creation options can control the behavior of the model. So, now is a good time to say a few more words about part design.

One modeling decision you often have to make involves the balance or trade-off between using a few very complicated features or many simple ones. You must consider the following when trying to put a lot of geometry into a single feature:

- How easy will it be to modify the part/feature later?
- If the geometry is very complex, it is generally easier to create a number of simpler features that would combine to give the same resulting geometry.
- Using more, but simpler, features generally will give you more flexibility.
- Having a higher feature count increases the need for a carefully managed parent/child network.
- If you plan to do some engineering analysis of the part, for example a finite element analysis, then minor features such as rounds, chamfers, small holes, etc., will only complicate the model, perhaps unnecessarily. They will also lead to increased modeling effort downstream. These features are normally added last. We saw in Lesson #5 how they can be temporarily excluded from the model (called *suppressing* the feature), as long as they are not references (parents) of other features.
- If the entire part is contained in a single feature, some major changes to the part may not be feasible using that feature.
- What is the design intent of each feature? How should each feature be related to other features (via the parent/child relations)? Don't set up unnecessary interdependencies between features that will restrict your freedom of modification later.
- You must be very careful with references. Sometimes these are essential elements of the design intent; sometimes you will fall into the trap of using a reference as a convenience when setting up a sketch, where this is not in the design intent. If you try to modify the feature later, you may find that the reference will get in the way.

When creating the patterns and copies, we discovered the ways that duplicated features could be modified, either during feature creation or after the fact. We also saw some of the ramifications of feature order in the model.

These considerations should be kept in mind as you plan the creation of each new part. It is likely that there are many ways in which to set up the part, and each will have different advantages and disadvantages depending on your goals. The more you know about the Pro/E tools, and the more practice you get, the better you will be able to make good decisions about part design. Good planning will lead to an easier task of part creation and make it easier to modify the geometry of the part later. Like most design tasks, the model design is subject to some iteration. We discussed in Lesson #5 some of the tools that Pro/E provides (the three R's) to allow you to change the structure of your model if it becomes necessary or to recover from modeling errors.

Most importantly, since design is increasingly becoming a group activity, make sure your model will be easy for someone else on your design team to understand. They may have to make modifications while you are away on vacation and you want them to be happy with you when you get back!

In the next lesson we will see how to create an engineering drawing from a Pro/E part. This will include view layout, section and detail views, and dimensioning. We will also create a couple of parts that will be needed in our assembly in the following two lessons.

Questions for Review

1.   When creating a revolved protrusion, does the sketch have to be open or closed or can it be either?
2.   When creating a revolved cut, does the sketch have to be open or closed or either?
3.   What essential element is common to all revolved features?
4.   Suppose you are creating a revolved protrusion and you align a vertex of the sketch with an existing feature surface. What happens if you try to create a 360 degree revolve and the aligning surface doesn't exist for the full revolution?
5.   What is the first feature in a pattern called?
6.   What is meant by a "radial" hole?
7.   How do you turn off the note that comes with a standard hole?
8.   How can you tell if a hole has a thread on it?
9.   What dimensions are available for patterning a feature?
10.   How could you create a spiral pattern of holes?
11.   How could you create a radial pattern of rectangular cuts so that the orientation of the cuts (a) changes with each instance to stay aligned with a radial line or (b) stays constant relative to the fixed datums?
12.   What is the difference between independent and dependent copies?
13.   What is the easiest way to create a pattern of several related features?
14.   Is it possible to create a copy that is translated and rotated at the same time?
15.   Comment on the rule of thumb in solid modeling: "Add material first, subtract material last." Do you think this would generally lead to good modeling practice?
16.   What happens if you try to mirror one instance in a patterned feature?
17.   What happens if you try to pattern a feature created using a make datum as a sketching plane?
18.   What are the available variations for selecting references for specifying the translation direction and distance when making a copy?
19.   Are there any dimensions of a pattern leader you cannot increment in the pattern?
20.   Can you make a pattern of patterns? A group of patterns? A pattern of groups?

Exercises

Here are some parts to practice the features you have learned in this lesson.

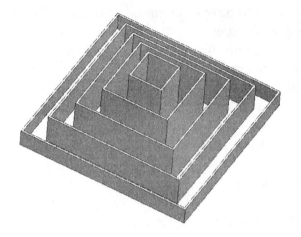

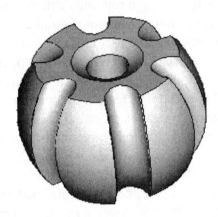

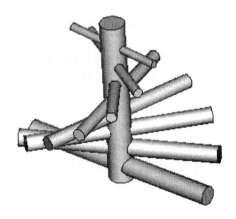

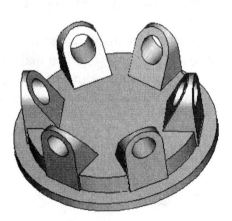

Project

This is the most complex project part. All dimensions are in millimeters. Some dimensions may be missing (because of implicit Sketcher rules or because the figures get too busy!). You can exercise some poetic license here and make a reasonable estimate for these. The important thing is that the assembly should fit together. If the geometry is confusing, visit the SDC web site for a VRML model of this part.

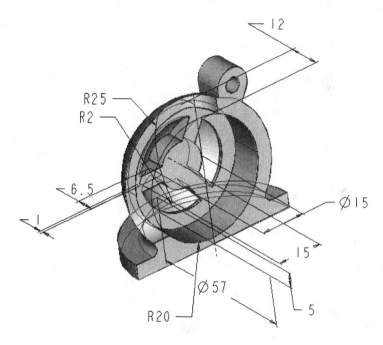

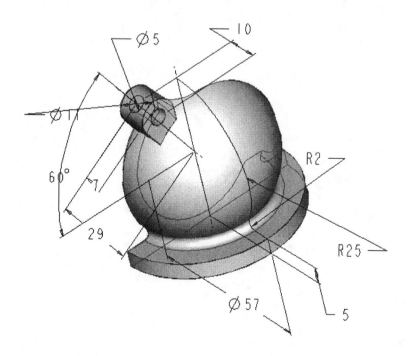

This page left blank.

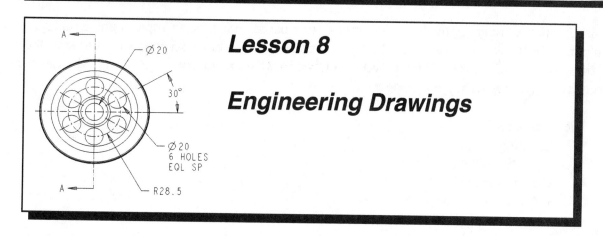

Synopsis

Creating dimensioned engineering drawings. Changing model units. View selection, orientation, and layout, section and detail views, dimensioning and detailing. Using a drawing template. Notes and parameters.

Overview of this Lesson

The primary form of design documentation is the engineering drawing. The drawing must contain complete and unambiguous information about the part geometry and size, plus information on part material, surface finish, manufacturing notes, and so on. Over the years, the layout and practices used in engineering drawings have become standardized. This makes it easier for anyone to read the drawing, once they know what the standards are. Fortunately, Pro/E makes creating drawings relatively easy. You will find that all the standard practices are basically built-in - if you accept the default action for commands, by and large the drawing will be satisfactory. There are a number of commands we will see that will improve the "cosmetics" of the drawing.

Doing drawings with Pro/E lets you concentrate a lot more on *what* to show in the drawing instead of dealing with the drudgery involved in *how* to show it. For example, when creating views of an object it is virtually impossible for Pro/E to create views of a part that is not physically realizable - we don't have to worry about any 3-pronged blivots (see the introduction). Even if the shape is possible, we don't want any visible or hidden lines in the wrong position (or missing entirely) on the drawing. This kind of mistake is easy to do with 2D CAD ("D" as in drafting) programs. The Pro/E solid model contains all necessary and sufficient information in order to define the part geometry. Therefore, by getting all this information into the drawing, it is very difficult to create a drawing with inaccurate information. That is not to say that creating drawings is totally automatic. For example, remember that when Pro/E interprets a sketch it fires a number of internal rules to solve the geometry. These rules are not indicated on the final drawing, and it may be necessary to augment the dimensions placed by Pro/E in order to complete the drawing to an acceptable industry standard. It is also possible, and sometimes necessary, to add additional graphical elements or notes to the drawing.

In this lesson, we are going to create drawings of two parts: an L-bracket support and a pulley. Both parts will be used in a subsequent lesson on creating assemblies, so don't forget to save the part files. We will also discover the power of bidirectional associativity, mentioned in the tutorial introduction. Here is what's on the agenda:

1.   The L-Bracket
     ▸   Creating the part
         •   changing part units
     ▸   Creating the drawing
         •   selecting the sheet
         •   creating the views
         •   adding dimensions
         •   cosmetic changes
         •   adding a note
     ▸   Changing the part/drawing - exploring associativity
     ▸   Sending the drawing to the printer
     ▸   Using a drawing template
2.   The Pulley
     ▸   Creating the part
     ▸   Creating the drawing
         •   selecting the sheet
         •   creating a section view
         •   creating a detailed view
         •   adding dimensions
         •   cosmetic changes
         •   using parameters in notes
         •   drawing options
         •   creating dimensions

Since this is the only lesson on drawings in this tutorial, we will only have time to cover the basics. Even at that, this is a long lesson. As usual, there are some Questions for Review, Exercises, and some Project parts at the end.

The L-Bracket

Creating the Part

First, we'll create the part shown in the figure at the right. Call this part **lbrack** and use the default template for a solid part. (We're going to change the part units in a minute.) Study this figure carefully. When you create the part, make sure that the back surface of the vertical leg is aligned with **FRONT**, the lower surface of the horizontal leg is aligned with **TOP**, and the vertical plane of symmetry is **RIGHT**. An obvious choice for the base

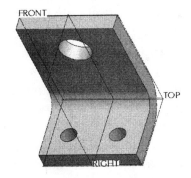

Figure 1 The L-bracket part

feature is a both-sides solid protrusion in the shape of a backwards "L" (as seen from the right side) sketched on **RIGHT**. This will allow us to mirror the second small hole and align the larger one with **RIGHT**.

The dimensions for the bracket are shown in Figure 2. Notice the dimensioning scheme for the holes.

Changing Part Units

Note that the units are given in millimeters, whereas in a standard Pro/E installation, the default template contains units of inches. This is a common "oops" when creating a model, since the units are not topmost in our mind when we first start the part (or when you inherit a model from another source). Here's how to change the part units. Select (from the pull-down menu)

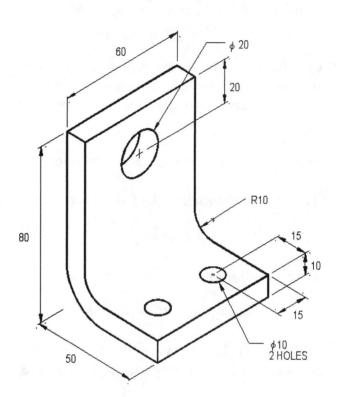

Edit > Set Up > Units

Figure 2 The L-bracket part dimensions

The **Units Manager** window opens, as shown in Figure 3. This lists the common unit systems in Pro/E (and its companion Pro/MECHANICA used for finite element analysis). The current units are indicated by the arrow pointer. Select the line containing the unit system

millimeter Newton Second

and then *Set*. A warning dialog opens. When you change the units of a model, you have two options that will affect all linear dimensions:

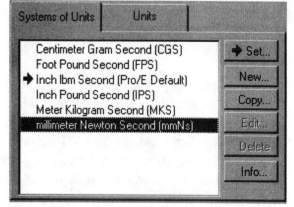

Figure 3 The Units Manager window

Convert existing numbers (Same Size) - This leaves the model the same real size as the original. For example, a 10 inch long bar will be converted to a 254mm long bar. The dimension number changes.

Interpret existing numbers (Same Dims) - This keeps the dimension numbers the same, but interprets them in the new units. In our example, the 10 inch long bar becomes a 10mm long bar.

Managing units is especially important if you are going to produce an assembly of parts, as we will do in the next two lessons. It is also critical to be aware of units when you are working in a design group, since some people may be working in inches while others are in millimeters. Parts downloaded from the web also come in all varieties. You may be aware of some classic blunders that have occurred over the years due to mix-ups in the interpretation of units.

If you have used the dimension values in the figure above, then you want to pick the second option here (*Interpret existing numbers*) and select *OK*. *Close* the Units Manager window. When this is applied, double-click on the protrusion to verify that the dimension numbers haven't change.

Don't forget to save the part! We are now ready to create the drawing.

Creating the Drawing of the L-Bracket

① Create the Drawing Sheet

Select the following:

> *File > New > Drawing | [lbrack]*

IMPORTANT: Deselect the option **Use default template**. We will deal with drawing templates a bit later. Select *OK*.

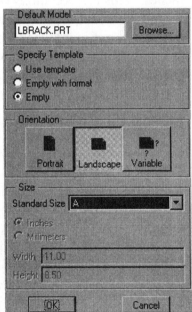

The **New Drawing** menu will open up as shown at the right. Note the currently active part is automatically selected as the drawing model. Keep the defaults for the template (**Empty**) and orientation (**Landscape**), but change the Standard Size option to **A** (8-1/2" by 11" in landscape mode). When this window is complete, accept the entries with *OK*.

A new window will open up with the title *LBRACK (Active)*. This will overlap or cover the part window, which is still open but pushed to the back. You can switch back and forth between the part and drawing windows using *Window* (in the pull-down menu). Current windows are listed at the bottom of the menu. If

Figure 4 The New Drawing menu

several windows are in view, only one of them will be active at a time (indicated by the word **Active** in the title).

In the drawing window, the Navigator displays the layer information for the drawing. Layers are a way to organize drawing information. We have lots to learn without dealing with layers, so you can close the Navigator (we will accept the default for anything that involves drawing layers, and won't be discussing them further).

Let's have a look at the user interface here, since quite a few things have changed. As usual when confronted with new menus for the first time, browse through the menu choices, paying attention to the message line at the bottom of the screen and the button tool tips.

Some new toolbar buttons have been added at the top and on the right. The top toolbar is quite extensive. It contains most of the basic commands for creating and managing drawings. In this lesson, we will only be using a few of them. Move the cursor over the buttons and observe the tool tip description (or message window). Despite the appearance of so many commands, creating entities with these drawing tools will not be a significant amount of work in Pro/E - almost all of the drawing will be done for you. The toolbar on the right contains drawing commands (as in creating lines, arcs, etc.). Some of these buttons look the same as those in Sketcher and perform a similar function. We will actually not be needing any of those commands here, either. Several of the pull-down menus at the very top have also been changed, or have new contents. Check them out. Finally, some new information is shown across the bottom of the graphics area.

② **Adding Views**

In the pull-down menu, select (most of these selections are defaults; remember you can accept an entire menu with middle-click)

> *Insert > Drawing View* (or use the top toolbar button *Add View* ⊞)
> *General | Full View | No Xsec | No Scale | Done*

Read the message window. The view we will place first will be our primary view. It will be the front view of the part, so select a center point a bit left and below the center of the sheet, as shown in Figure 5.

A drawing scale is set automatically, in this case it is 1.0 as shown in the bottom line in the graphics window. We can change that to a better value if required. So far, we have just selected the placement of the view. Now we want to reorient the part to get the proper front view. We do this by telling which surfaces or references in the model face which directions on the drawing. We will use the datum planes, although planar surfaces of the part could also be used. This is the same procedure as creating named and saved model views. Select the following in the **ORIENTATION** window (Figure 6):

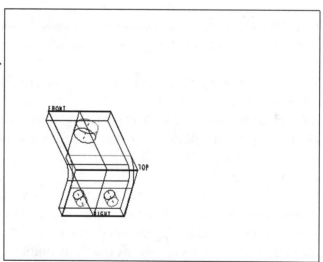

Figure 5 Placing the primary view

For Reference 1:
> *Front* > select the *FRONT* datum

For Reference 2:
> *Right* > select the *RIGHT* datum
> *OK*

Turn off the datum plane display. Observe the appearance of the tangent lines in the rounded corner. In the Environment (*Tools > Environment*) select *Tangent Edges (No Display) > OK*, then *Repaint*. Your drawing should look like Figure 7.

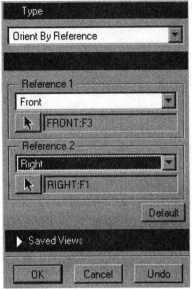

Figure 6 Orienting the view

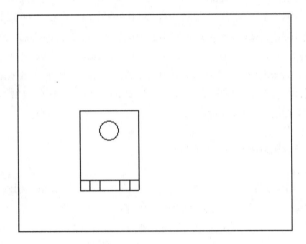

Figure 7 Primary view placed and oriented

Now we want to add the right and top views. These can be projected by Pro/E off the front view as follows. Use the *Add View* button (or select *Insert > Drawing View*), then select:

> *Projection | Full View | No Xsec | No Scale | Done*

then click on the drawing sheet to the right of the front view. Voilà! The right side view appears. Repeat the command and click above the front view to get the top view. This is easy! (But it gets even easier later.)

Notice that the dynamic mouse commands (you will primarily be using zoom and pan) are both on the middle mouse button. The difference is that the middle button by itself controls pan (rather than spin as in 3D mode). The scroll wheel (or CTRL-middle) controls zoom. Also, there is no spin center but *View Mode* is still available. There are other view control buttons in the top toolbar (*Zoom In* and *Zoom Out*, *Refit*).

If you don't like the spacing of your views, you can easily move them. By default, views are locked in place where you created them. In order to move a view, this lock must be turned off. You can do this in any of three ways:

- turn off the *Lock View* button in the top toolbar
or - select *Tools > Environment*, and turn off *Lock View Movement*
or - select a view, then in the RMB pop-up, turn off *Lock View Movement*

When you move views, Pro/E will ensure that your views stay aligned. Select the right side view - it will be surrounded by a red border and have a drag handle at the center. Use the drag handle to move the view. Try to move the view up, down, left, and right on the screen (you can't move up or down since the view must align with the front view since it is a projection). Left-click again to drop the view at the new location. Try moving the top view. Finally, try moving the front view. You should see the other views move to maintain the correct orthographic alignment.

Click the **left**-mouse button on an open area of the screen (ie not on another view) when you are finished moving the views to turn off view selection.

Let's add a fourth view that shows the part in 3D. Note that this is not a projected view but a general one. We'll scale this one down to half size. Select the *Add View* button, then

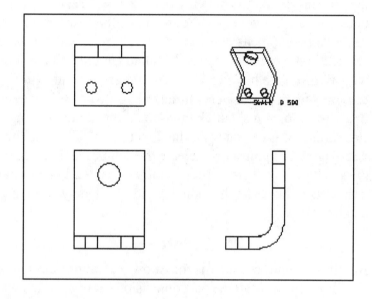

Figure 8 All views placed for L-brack

General | Full View
No Xsec | Scale | Done

and click in the upper right area of the drawing. Enter the scale factor *0.5*. Leave the part in the default orientation by selecting *Default >*
OK in the **Orientation** window. Your screen should look like Figure 8.

③ **Adding Dimensioning Detail**

In the pull-down menu select (or select the *Show/Erase* button)

View > Show and Erase

A new window opens with a number of detailing types and options as shown in Figure 9. Move the mouse cursor over each of the buttons in the **Type** area - the message window (and a pop-up) will show you what the button does. Select

Dimension (the top left button)
Show By (Part)
Preview > With Preview (should be checked)
Show All

Because we selected *Show All* (which is a potentially hazardous thing to do - think of a part or assembly that might have hundreds of dimensions!), confirmation will be requested (select *Yes*). All of the part dimensions used to create the model will be put up on the display in dark red. We have several options for dealing with these: erase them all, keep them all, or just select the individual ones we want to keep or remove. There aren't too many dimensions in this drawing so select

Accept All > Close

in the **Show/Erase** window. Click anywhere on the drawing. The dimensions will change to yellow.

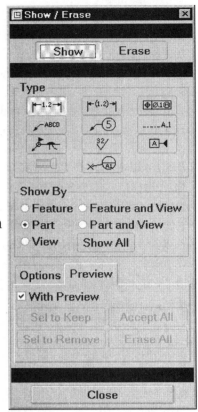

Figure 9 The Show/Erase menu

Take a moment to think back to how you created the part. The dimensions put on the drawing using **Show/Erase** are exactly the ones you used in your sketches to create the features. These are called *Shown* dimensions. Another type of dimension can be created in the drawing, naturally called *Created* dimensions, which we will discuss a bit later. **The lesson here is to use the dimensions in feature creation that you want to appear on the drawing**[1]. So, you should know something about drawing standards and how you want the dimensions laid out in the drawing *before* you start to create the solid model - a point often missed by many. This is often a point of contention between designers (who make the models) and detailers (who create the detailed drawings).

Another thing to consider is the **Show By** option selected above. We chose to show all the part dimensions at once. This was all right for this simple part since there were not too many dimensions to deal with. For more complicated parts, you might like to show the dimensions by individual feature, all dimensions in a given view, or a specific feature in a chosen view. Some experience with these options is necessary to make good choices here, otherwise you'll spend a lot of unproductive time cleaning up the drawing.

In the next drawing, we will see a more systematic way of adding the dimensions rather than *Show All*.

④ **Dimension Cosmetics**

Although all the dimensions are now on the drawing, and Pro/E does the best job it can to determine where to place the dimensions (which view, and so on), there is a lot we may need to do to improve their placement and appearance. For example, some of the dimensions may be a little bit crowded. To fix this, select the following (or use the *Cleanup* button)

Edit > Cleanup > Dimensions

---

[1] This is a good topic for a philosophical discussion. Some users differentiate between "design" intent (dimensions used to create the computer model) and "manufacturing" intent (dimensions appearing on the drawing), maintaining that sometimes these are different.

This opens the window shown in Figure 10. We have to identify which dimensions we want cleaned. Hold down the left mouse button and draw a selection box around the entire drawing, then select **OK**[2]. The number of dimensions affected will appear at the top of the **Clean Dimensions** window. Accept the default distances for the offsets (the 0.5 is the spacing in real inches from the edge of the part to the first dimension line, the 0.375 is the offset between parallel dimension lines - these are drawing standards). Then pick on the *Apply* button. All the dimensions should spread out and appear in dark red. The dashed gray lines are called the snap lines. As you proceed to modify the drawing layout, the dimensions will snap to these locations to help you maintain the spacings set in *Clean Dimensions*. These snap lines are a convenience and will not be printed with the drawing.

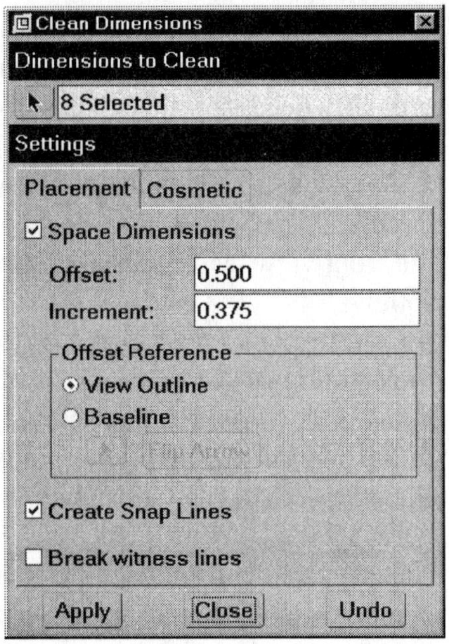

Figure 10 Cleaning up dimensions

Depending on your view placement and dimensioning scheme, Pro/E might have some trouble with dimension placement (for example, too little room between views). Look in the message window for error messages and warnings. Close the **Clean Dimensions** window. All dimensions should now be in yellow.

The drawing should look something like Figure 11 (your dimensioning scheme may be slightly different from this, depending on how you created your model):

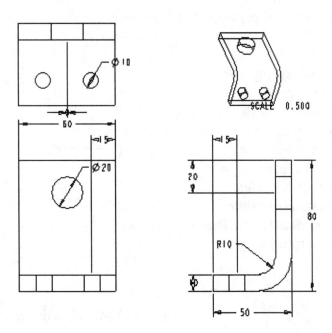

Figure 11 Dimensions placed and cleaned

---

[2] This is one of the very few times in Pro/E when selected items don't highlight!

There is a lot more we can do to modify the display "esthetics" of the dimensioning detail. Some of the dimension placement locations chosen by Pro/E may need to be touched up a little. It is probably necessary to switch some of the dimensions to a different view, and you may want to modify spacing and location of dimensions on views, direction of dimension arrows, and so on. For example, the location dimensions for all the holes should be on the view that shows the circular shape of the hole. For the two small holes, this is the top view. For the large hole, this is the front view. Most of these cosmetic modifications can be made using the mouse buttons as follows.

Pick (left click) on one of the dimensions you want to modify. For example, you might select the dimension giving the thickness of the plate as shown in the right view. (See Figure 12). Note the small square "handles" on the dimension components. Left-click on any handle in order to drag it to the desired position. If you select the handle directly under the dimension value, you can move it practically anywhere. The extension lines and arrows will automatically follow.

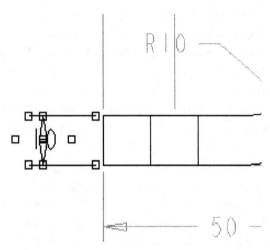

Figure 12 "Handles" for modifying dimension cosmetics

While dragging this around, if you want to flip the dimension arrows (ie. put them inside/outside the extension lines), just ***Right-click***. Notice the effect of the snap lines.

When the dimension is where you want it, ***left-click*** to drop. You can continue to left-click on the handles to move the dimension, extension lines, dimension line, and arrows until you get exactly the appearance you want. ***To accept the final placement and format, click the left mouse button somewhere else on the screen or select another detail item.*** The modified dimension will turn yellow.

The final configuration might look something like Figure 13.

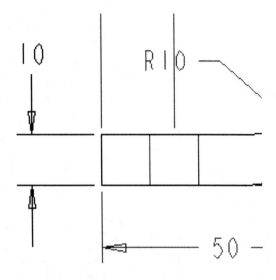

Figure 13 Modified dimension cosmetics

To modify more dimensions, continue the sequence:

- ▸ left-click on a dimension,
- ▸ drag on the handles as desired using left-mouse,

> ‣ right-click to flip arrows if desired,
> ‣ left-click to accept

until you are satisfied with the layout. If you want to move a dimension to another view, after you have picked out the dimension, hold down the right mouse button and select *Move Item to View* from the pop-up menu, then left-click on the desired new view. You can also use (in the pull-down menu) *Edit > Move Item to View*.

Also available in the pop-up menu are a number of other cosmetic modification commands. The major ones are fairly self-explanatory, however one is mentioned here because of an important concept you need to know (you can consult the on-line help for details about the others):

Erase

removes detail items from the drawing. Note that this is not the same as *Delete*. With erase, the dimension still stays with the model, it just isn't displayed. It makes sense that a shown dimension that is part of the model cannot be deleted. However, you can create dimensions that *can* be deleted, since they are not necessary parts of the model.

Try to lay out all the dimensions so that your drawing looks similar to Figure 14. The dashed offset (snap) lines created when we cleaned the dimensions can be removed by selecting *Delete* from the pop-up menu. You don't really need to do this for hard copy, since Pro/E will not print snap lines.

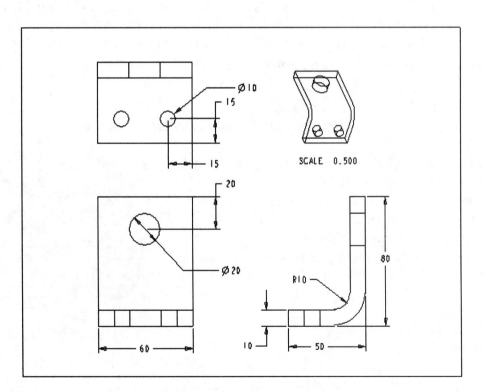

Figure 14 L-bracket final drawing

Do not be concerned at this time if the yellow extension lines are touching or crossing the model. As you probably know, this is a "no-no" in engineering drawings. Pro/E will clean up the extension lines when a hard copy is generated. Pro/E will also look after all the line weights and line styles (for visible and hidden lines, center lines, dimension and extension lines, and so on) according to standard engineering drawing practice.

⑤ Setting View Display Mode

The default view display (hidden line, wireframe, etc) is determined by the view display of the part in the model window. Of course, we can't use shaded images in a drawing. Thus, it is possible that your drawing is displaying all edges as solid lines (hidden line and visible lines the same). To make sure hidden lines are treated correctly, select all four views (using the CTRL key), then in the RMB pop-up menu, select **Properties**. (You can also do this one view at a time by double clicking on it.) In the **View Modify** menu select

>   *View Disp*
>   *Hidden Line | No Qlt HLR | No Disp Tan | Drawing Color | Done*

then press *Repaint*. These view display settings are now fixed properties of the views and not affected by the top toolbar buttons.

⑥ Creating a Note

Let's add a short note on the drawing (we will talk about title blocks in the next section). You may have to move the other views up a bit to fit this in (you can do that after the note is created, if necessary). Select in the pull-down menu (or use the *Create Note* button [A≡])

>   *Insert > Note*
>   *No Leader | Enter | Horizontal | Standard | Default | Make Note*

Select a location a little below the right side view. A small **Text Symbol** window opens from which you can select special characters to insert in the note. Normal characters are just typed in (see the message window). Pressing the enter key will advance you to the next line of text in the note. Pressing the enter key on a blank line will complete the note. Type in something like the following:

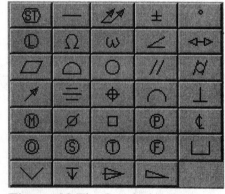

Figure 15 The Symbol Palette

ALL DIMENSIONS IN mm
Drawn by Art O'Graphic
22 Feb 2003

Select *Done/Return*. If you left click on the note, you can move it around. If it is selected, you will find the *Text Style* command in the RMB pop-up. This lets you change the text font. To edit the text itself, select the note and then pick (in the pull-down menu) *Edit > Properties*.

Save the drawing using the default filename; Pro/E will automatically append a *drw* extension to the file name.

>   *File > Save*

Exploring Associativity

One of the most powerful features of Pro/E is its ability to connect the part model and the drawing. Here is a scenario where this is very useful.

It's late Friday afternoon and your boss has just reviewed the design and drawings of the L-bracket, and has decided that a few changes are needed as follows (before you go home!)

- ◆   the height must be increased to 100 mm
- ◆   the diameter of the large hole must be changed to 30 mm
- ◆   the top of the bracket must be rounded in an arc concentric with the large hole
- ◆   the manufacturing group wants the drawing to show the height of the large hole off the bottom of the part, which should be 70mm

Hmmmm... You could do this by going back to the part and modifying/changing. BUT..there is an easier way! To really see the power of what you are about to do, resize the drawing and part windows so that both are visible.

Make sure the **DRAWING** window is active. If not, click on the drawing window and select *Window > Activate* in the pull-down menu.

Double-click on the diameter dimension of the large hole. Enter a new value of *30*. Click somewhere off the dimension and it will show in green. Double-click on the height dimension and change it to *100*. Now, in the top toolbar menu, select

>   *Regenerate*

The drawing should change to show the new geometry. Even better, go to the part window and activate it (use CTRL-A). *Repaint* the screen. It also shows the new geometry. Double click on the protrusion and change the width of the bracket from 60 to *80*, then *Regenerate*. Change back to the drawing window and activate it - it shows the new shape too. These actions show that there is a *bidirectional* link between the drawing and the part. If changes are made to any item, the other is automatically updated. The same holds true when you deal with assemblies of parts, and drawings of those assemblies.

Consider what Pro/E needs to be able to accomplish this associativity between part and drawing. In order to update the part when a dimension is changed on the drawing, then the part itself must be "in session" whenever the drawing is[3].   Therefore, to bring up the drawing (read it from disk), the part file must also be available (on disk or already in session). The part will be brought into

---

[3] Check out *Info > Session Info > Object List*

session, if necessary, even if it is not displayed in its own window.   Furthermore, when read from disk, the part file must be in the same location relative to the drawing file as it was when the drawing was created (usually the same directory) - otherwise Pro/E doesn't know where it is and cannot load it. You cannot load the drawing as an independent object. The same holds true for assemblies: when you want to work on an assembly, all its constituent part files must be available to be read in as required. You can see that for complicated projects, file management might become an issue.

Consider also that if you change a dimension on the drawing, this makes a change in the part file loaded in session. It does not immediately change the part file on disk. To make the change "stick" you have to save the part file! The same holds for some other drawing entities like section definitions[4].

Before we forget, change the width of the bracket back to **60**. In the drawing, select the dimension to highlight it, right click to bring up the pop-up menu, select *Edit Value*, enter in the new value **60**, select *Regenerate*.

One thing we can't do in the **DRAWING** window is change the basic features of the part (like creating new solid features, or changing feature references). For that you have to go back to the **PART** window. Do that now, so that we can add the cut to round off the top of the bracket.

First, if necessary, *Reroute* the large hole (select it and in the RMB menu select *Edit References*) so that the horizontal dimension reference is **TOP** instead of the upper surface of the bracket. The distance above this reference should be **70**. If the hole disappears off the bottom of the part (the axis is still visible), modify its dimension value to -70. If you scroll back a few lines in the message window[5], you will see a warning that was produced when the hole was regenerated (something like "Hole is entirely outside the model").

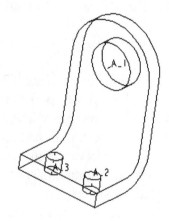

Now create a circular arc cut, concentric with the large hole and aligned with the left and right sides of the bracket. The part should look like Figure 16 when you are finished. Don't forget to save the new part.

Now we have to touch up the drawing a little. Change back over to the drawing window. The new circular arc should be shown there.

First, the drawing scale is a little too big for the sheet. Click on the *Scale* value shown on the

Figure 16 The modified L-bracket

bottom line of the graphics window. In the RMB menu, select *Edit Value* and enter *0.8*. You might like to reposition the views. Notice that this change in scale affects the orthographic views (that are using the sheet scale) and not the 3D general view (that has its own scale).

Next, you may note that the large arc isn't dimensioned. Actually, a dimension isn't needed for the arc since we know the block width. And anyway, because of the way the feature was created, it has no dimensions in the model, so there is nothing to *Show*! Since the dimension is not strictly necessary, we will create a *reference dimension* in the drawing. Select (in the pull-down menu)

Insert > Reference Dimension > New References > On Entity

and then left click on the arc in the front view. Use the MMB to place the dimension. Select *Return* in the **ATTACH TYPE** menu. You might like to clean up the dimension cosmetics a bit. In Pro/E language, the dimension we just added is called a *created* dimension (as opposed to a *shown* dimension). These do not have to be reference dimensions.

Let's add the axis lines for the holes. Select the *Show/Erase* button and pick the following:

> *Axis* (middle button in right column)
> *Part > Show All > Accept All > Close*

You can get rid of the axis labels **A_1**, **A_2**, etc. by turning off the axis display using one of the short-cut buttons, then *Repaint*. This leaves the axes but removes the labels when you are in drawing mode.

You should also change the text in the note. Left click on the note, then in the pull-down menu select *Edit > Properties*. The text will appear in the text editor window. Change the first line to something like

SCALE 0.8, DIMENSIONS IN mm

Select *OK* when you are finished and don't forget to save the drawing.

Getting Hard Copy of the Drawing

Obtaining hard copy depends on the details of your local installation. See your system administrator for information on this. However, there are two possible ways that might work.

If you are running under Windows with an attached printer, try this:

File > Print

or use the *Print* shortcut button. In the **Destination** field of the new window, select *MS Printer Manager*, then

Configure > Model > Plot (Full Plot) > OK
OK

This should bring up your normal Windows print control dialog. Use it as you usually would to select the printer and printer properties (quality, speed, color, page size, etc). Some experimentation may be required here to get margins, orientation, and so on set just right.

If you do not have a plotter attached directly (or wish to archive the drawing image for use in another program), obtaining a hard copy of the drawing is a two-step process. First, we create a postscript-format file of the drawing, then copy the file to a postscript-capable printer (or other application that understands postscript[6]). Try this:

File > Print

Click the button to the right of the **Destination** field, select *Generic Postscript*, then

Configure > Model > Plot (Full Plot) > OK
*To File (*and deselect *To Printer)*
OK

A dialog box will open asking you for the name and path of the file. The default will be *lbrack.plt* in the current working directory (unless this has been over-ridden by your system administrator). Click *OK* to generate the file. Once you have a postscript file of the drawing, there are a number of ways to obtain hard copy. You will need access to a postscript-capable printer and you may have to find out how to transfer the file from your Pro/E computer/directory to the printer. Generally, once you have the file on a computer connected directly to a postscript-capable printer, you only need to copy the file directly to the printer. See your system administrator for further information.

Using Drawing Templates

For our first drawing, we did a number of operations manually. Many of these are common to all part drawings. Fortunately, there is a way to do much of this tedious drawing creation automatically.

First, make sure the current drawing has been saved, then erase it with *File > Erase > Current*. Note that this does not delete the drawing from your hard disk but just removes it from the current session (takes it out of memory). You should be back in the part window.

Create a new drawing called *lbrack2*. Once again, deselect the option beside **Use default template** (we want to pick our own) and select *OK*. In the **New Drawing** dialog window, check the button beside *Use template* and select an **A** sized drawing by picking *a_drawing* in the **Template** area. This does the following:

---

[6] Converting drawings from postscript to *pdf* format is common - these are easily shared but cannot be marked up by the user.

- creates the drawing sheet (A size)
- orients the model
- places the standard views (top, front, right) for a multiview drawing
- scales the views to give you room for detailing

When you select **OK** and enter the drawing window, everything should be set up for you as shown in Figure 17.

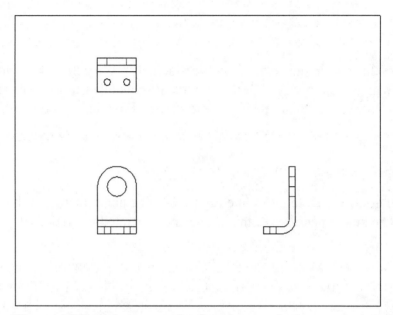

Figure 17 Drawing created using the template

How does Pro/E know what standard views you wanted? The views that are shown on the drawing are determined by the layout of the template[7]. The template refers to standard views embedded in the part model, that was created with the part template. The drawing views are based on the default datum planes TOP, FRONT, and RIGHT and the associated Saved Views. The orientation of the part in the drawing is therefore determined by how we orient the geometry of the part relative to the datums. If your part is upside down in the model, then the drawing views will be upside down too. Another good reason to plan ahead!

Now, there may be a good reason to have the orientation of the part different in the model than in the drawing. If you still want to use the part and drawing templates, here is how to reorient the drawing views created automatically. Double-click on the front (primary) view. In the VIEW MODIFY menu, select **Reorient**. The other views will be surrounded by magenta boxes, and in the message window you are asked whether you want these reoriented as well (to maintain projection). Select the **Yes** button or type in a "*y*". The **Orientation** window appears. In the **Saved Views** region, select the **Left** view and then the **Set** button. The primary view reorients to the predefined **LEFT** view of the model, and the other drawing views also reorient to suit. Change back to the original **FRONT** view and then use **OK** to close this dialog window.

---

[7] Creating your own templates is discussed in the *Advanced Tutorial*.

With the views created, go ahead and finish detailing the drawing for practice. Try to do this on your own, but refer back to our previous procedures if necessary.

Now, on to the second part. This will require creating a section view, a detail view, and a quick look at drawing title blocks and customization tools.

The Pulley

We're going to use this part in the next lesson (on assembly). We will make it now to see how to create a drawing with a section view. We'll also look at some other things we can do when creating drawings, like setting up a title block and border. First, let's get on with the pulley model.

Creating the Pulley

The pulley we are going to create looks like Figure 18. The main interest in this part is the cross sectional shape. The key dimensions of this shape are illustrated in Figure 19.

Figure 18 The pulley

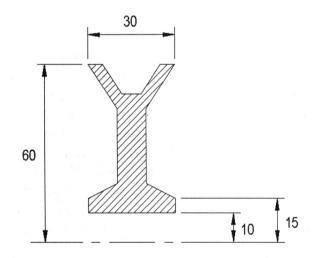

Figure 19 Pulley cross section

We could create the base feature as a single revolved protrusion. However, this single feature would be difficult to set up in the sketcher. Instead, we'll create the pulley using a number of features (we'll use about 12 in all, including the holes and rounds).

Start by creating a new part called *pulley* using the *mmns_solid_part* template. In the appropriate data fields, enter something like *[Tutorial pulley #1]* for the *Description* and your initials for *Modeled_by* parameters.

Create a circular disk (both sided protrusion off **FRONT**) aligned with the origin. Look ahead to Figure 27 or 30 to see why we want this orientation. The disk has a diameter of *120* and a thickness of *30*. The disk should look like Figure 20.

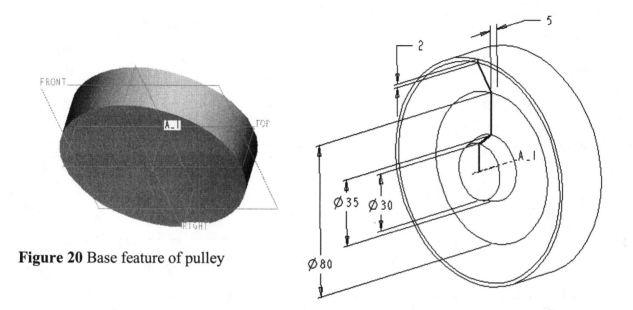

Figure 20 Base feature of pulley

Figure 21 Revolved cut on one side of pulley

Now, create a 360 degree revolved cut on one side of the disk. The sketching plane is **RIGHT**. The dimensions are shown in Figure 21. The revolved cut can be mirrored through **FRONT**.

Now create the pulley groove around the outer circumference as another revolved cut. Just make a symmetrical 60° V-shaped groove as shown in Figure 22. The vertex at the bottom of the V aligns with **FRONT**.

Add a round at the bottom of the groove with a radius of *3*.

Add the central hole for the pulley axle. This can be created as a *coaxial hole* off **FRONT** with a diameter of *20.* The depth is *Thru All* in both directions. See Figure 23.

Now we'll create the pattern of holes arranged around the pulley. We start by creating the pattern leader. Again, use

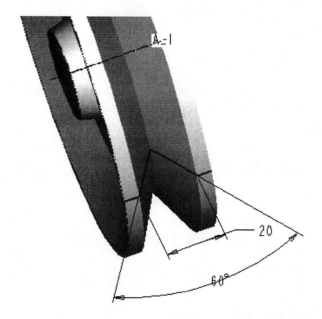

Figure 22 Revolved cut to make pulley groove

FRONT as the placement plane and go *Thru All* in both directions. Create the hole using the *radial* option (*28.5* from pulley axis). Measure the angle *30* from **TOP**. This is the angle that we will increment to make the pattern. See Figure 24.

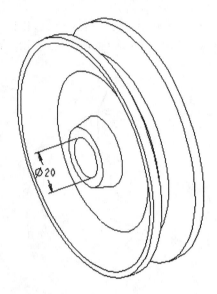

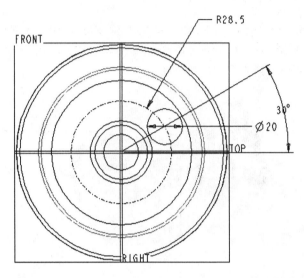

Figure 24 Hole Pattern leader

Figure 23 Central hole added to pulley

Create the pattern using the first hole as the leader. Increment the angular dimension by *60* and make a total of *6* holes.

As a final touch, add some *rounds* (radius *1*) to the outer edges as shown here. All four edges are in the same feature - don't create four separate rounds!

That completes the creation of the pulley. Before we go on to the drawing, don't forget to save the part.

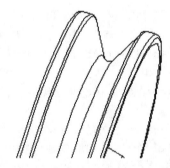

Figure 25 Rounds added to outer edges

Creating the Drawing

① **Selecting a Formatted Sheet**

For this drawing, we will use a pre-formatted sheet with a title block. Start a new drawing with

File > New > Drawing > [pulley]

You can leave the default template box checked. In the **New Drawing** window that opens up, select *Empty with format*. In the **Format** area, select *Browse* to find the path to the directory on your system that contains drawing formats. The default location is identified by **System Formats** in the **Look In** list at the top. We are looking for a file called **a.frm**. The default location is (for Windows systems with a "generic" Pro/E installation) **/ptc/proe2001/formats/a.frm**. If you can't find it, either consult your system administrator, or carry on without the format by canceling the command. In the **New Drawing** window, select *OK*.

Assuming you were able to load the format, the drawing window will open with an ANSI standard title block and border already drawn on the A-sized sheet as shown in Figure 26. You can close the Navigator.

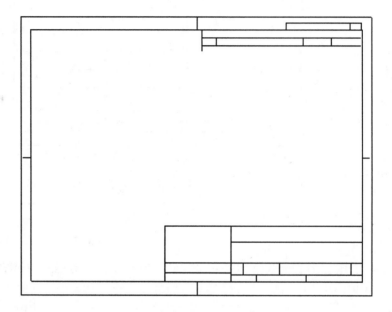

Figure 26 Formatted drawing sheet (A size)

② **Creating the Primary View**

Create a front view of the pulley showing its circular profile. Use the *Add View* button and select:

General | Full View | No Xsec | No Scale | Done

Click to the left of center of the sheet. In the **Orientation** window near the bottom click on the region labeled **Saved Views**. This opens a list of the views saved in the part template. Select

FRONT > Set > OK

You may have to change the sheet scale (indicated at the bottom of the graphics window) to *0.5*. Your screen should now look like Figure 27.

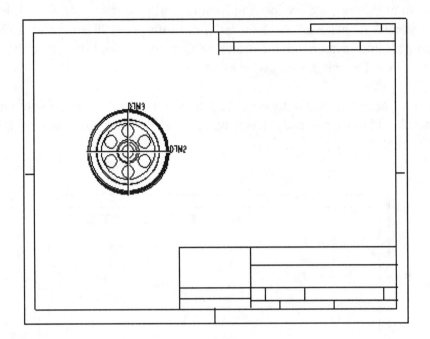

Figure 27 Primary view placed and oriented

③ Add a Full Section View

We will create a full section to the right of the primary view. To do this, we have to specify the type of view, the location of the view, where the section is to be taken, and on what view to indicate the section line. Use *Add View* again and select the following (follow the prompts in the message window while you do this):

Projection | Full View | Section | No Scale | Done

Then in the **XSEC TYPE** menu, select

Full | Total Xsec | Done

Pick a location on the sheet off to the right. Now we have to tell Pro/E what to call the view and where we want the section taken. In the **XSEC ENTER** menu select:

Create > Planar | Single | Done > [A]

Because of our last entry, our section will be identified as *Section A-A*. Now we specify the cutting plane for the section. We want to use a vertical plane through the pulley. If a datum plane doesn't exist for this, you could create a make datum. In our case, **RIGHT** will do just fine. Select it in the front view, then *OK*.

Read the message window. Pro/E is asking you on which view to put the cutting line ("view for the arrows"). Pick on the front view. We are finished with the datum planes, so you can turn them off now. Your drawing should look like Figure 28.

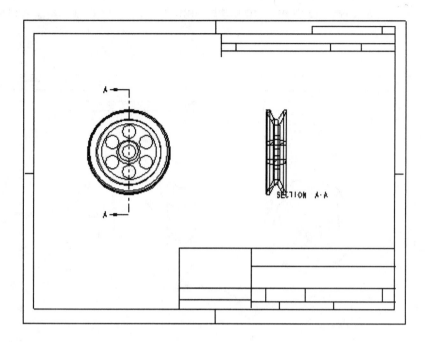

Figure 28 Section view placed

④ **Modify the Section View Display**

Section views generally do not show any hidden edges. Let's turn them off. Double-click on the section view at the right. In the VIEW MODIFY menu, select

> ***View Disp***
> ***No Hidden | No Disp Tan | Drawing Color | Done***

⑤ **Adding a Detail View**

We'll add a broken out detail view of the pulley groove. This will be useful for dimensioning and showing the rounds. We'll also draw this at twice the scale of the drawing. Select the ***Add View*** toolbar button, then pick

> ***Detailed | Full View | No Xsec | Scale | Done***

Read the message window prompts carefully as you do this. Pick a point on the drawing where there will be enough space for the view (we can always move the view later if this point doesn't work out). See Figure 30. At the prompt for the Scale, enter *1.0*. Now pick a point near the bottom of the pulley groove in the section view. We now want to indicate the area around the pick point to be included within the detailed view. As you click with the left mouse button, a red spline curve will be drawn. Make sure this encloses the groove (four points should be enough). When you have fully enclosed the area to be drawn, click with the middle mouse button. Enter the name of the view, *B* and select the ***Circle*** boundary type. A circle will appear roughly around the area you identified, and you can pick (left click) a location for a note to identify the circle. This can be moved later if required. You should now have a scaled-up detailed view something like Figure 29. You can move the views around by unlocking them and dragging them wherever you want.

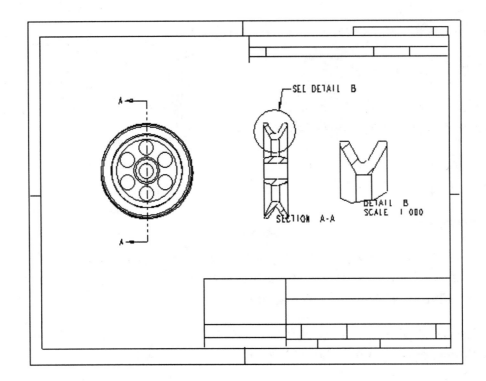

Figure 29 Detail view of section added

⑥ Adding Dimension Details

Instead of getting Pro/E to show us all the dimensions at once using **Show All**, we will be a little more selective since this part has quite a few dimensions. This will give us more control about initial placement of dimensions, which means fewer changes later (hopefully!). We are going to select features in the model tree, so open that now. (If the Layer display is still there, select **Show > Model Tree** in the Navigator window). In the model tree, select the base feature and in the RMB pop-up menu, select **Show Dimensions**. The dimensions for only the selected feature appear on the drawing (for this feature, look for the diameter and thickness). You can easily manage the cosmetic changes for these. Look ahead to Figure 30 for some drawing layout ideas.

Right click on the next feature listed in the model tree and then **Show Dimensions** in the RMB pop-up. Clean these up using the methods we saw previously (including **Cleanup Dimensions** if things get really messy!). Continue moving down the model tree, picking one feature (or more) at a time and Showing the dimensions. For the hole pattern, dimension the pattern leader and change(use **Edit > Properties**) the diameter dimension text as shown in Figure 30.

You may find that picking features out of the model tree (if you know exactly what they are in the drawing) is the easiest way to manage the creation of the drawing. In a complex model, it really helps if the features are all named *and* you have thought about how you want to lay out the dimensions before you start! Following standard drawing practice, Pro/E will place each necessary dimension only once (this will be important in a minute or two!), so if you want a dimension in a particular view you must either first create it there, or use *Move Item to View* later.

When you are finished adding detail, close the Navigator.

⑦ **Improving the Esthetics**

If you haven't already, use the drag handles and the right mouse button to modify/move the dimension details as required to get a better layout.

Let's change the crosshatch pattern in the section and detailed views. Select the hatch in the section view. The in the RMB pop-up menu select

> *Properties*

You can now change hatch spacing, angle, pattern, and so on. The *Retrieve* command lets you choose from standard hatching patterns for different materials (aluminum, iron, copper, steel, and so on). To change the spacing and angle, select

> *Spacing | Hatch > Overall | Half* (click twice)
> *Angle | Hatch > Overall | 30 > Done*

Note that the hatching changes in the detail view as well.

Next, we'll add all the centerlines for circular features. Select the *Show/Erase* button, then

> *Axis | Part | Show All*

Use *Select to Remove* to retain only the desired axes (there are a couple on the section view that we don't want shown). Turn off the axis labels using the *Datum Axis* display button.

If you have trouble removing the unwanted centerlines, select *Show/Erase* and pick the *Erase* tab. With the *Axis* button still selected, and the radio button checked for *Selected Items*, pick on the offending centerlines. Hold the CTRL key to pick more than one. You may find two centerlines on top of each other. When you are finished, close the Show/Erase window.

⑧ **Adding Notes with Parameters**

Finally, add some text to the title block. You can, of course, use notes to create plain text within the title box. You may want some notes to change if the model changes. You can do this with parameters. In the pull-down menu select (or use the Create Note button):

> *Insert > Note*
> *No Leader | Enter | Horizontal | Standard | Default | Make Note*

Do you remember entering a value for the parameter *DESCRIPTION* when creating the part using the template? The text was something like "Tutorial pulley #1". Pick a point in the appropriate cell in the title block (see Figure 30). Then type in the following text in the prompt area (without the square brackets):

[&description]

Press enter (twice) when you are finished. The value of the part parameter will appear at the insertion point - this is what the "&" symbol does when used with parameters. You can move the note to center it in the box. Put a note for the *MODELED_BY* parameter in another box in the title block using the text string

[Modeled by &modeled_by]

Notice that when you select the insertion point, all the dimensions in view are changed to their symbolic form. Try entering a note with the following text (observe the dimension symbol on your drawing for the diameter of the central hole in the pulley):

[Pulley shaft ⌀ &d15]

For the diameter symbol, use the symbol palette at the right. When you accept this, what happens to the dimension on the drawing? Why does this happen? In any case, it is not good practice to put part dimensions in the title block, so highlight this note and delete it. What happens to the hole dimension?

How could you enter the note to display the drawing scale?

⑨ Changing Drawing Options

It is likely that the centerlines on the hole pattern look slightly different from Figure 30. The display of these centerlines is controlled by a drawing option. Other options include text height, arrowhead size and style, tolerance display, and many more. To see the options, with nothing in the drawing selected use the RMB pop-up and select *Properties* (or in the pull-down menu select *Edit > Properties*). Then select

Drawing Options

This brings up a long list of options. They are sorted **By Category** (see the setting at the top of the window). Default values/settings are indicated with an asterisk "*". This information was read from a default file on your system when you created the drawing. Any values you change here will affect only the current drawing (unless you can change and over-write the system file) and be stored with it.

Browse down this list to get a feel for what you can do in terms of drawing customization.

About 2/3 of the way down the list (or sort the list alphabetically), look for the option

radial_pattern_axis_circle

and set it to *Yes*. Then select *Add/Change > Apply > Close > Repaint*. The centerline layout should now be circular and radial.

Your final drawing should look something like Figure 30. Here is a test of your drawing-reading abilities: what is the missing dimension in Figure 30?

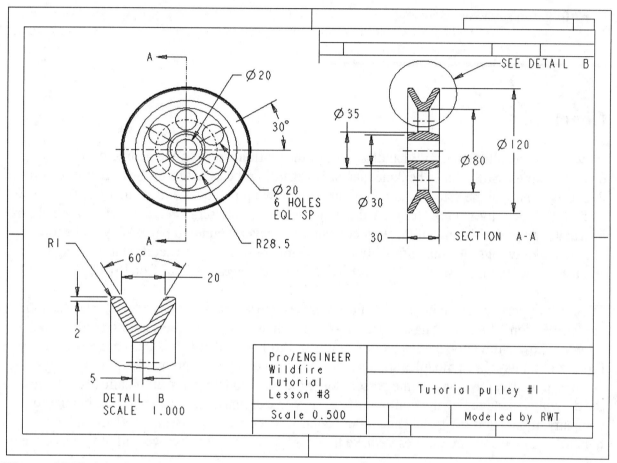

Figure 30 Finished pulley drawing. Can you find the missing dimension?

⑩ Creating Dimensions

As mentioned earlier, the dimensions placed by Pro/E are the ones used explicitly to construct the model. From time to time, you may have to add some dimensions manually. You can do this using

Insert > Dimension > New References

The dimensioning tools available are fairly self-explanatory and will be easy to pick up by anyone who has done 2D CAD. Two things must be remembered however. First, dimensions that you create this way can be deleted. Second, the dimensions you create cannot be used to drive the geometry - they are strictly lines on the drawing. These are called "*driven*" dimensions and cannot be modified in the drawing (but will change if the geometry of the part changes).

You can also use the Sketcher-like tools to add entities to the drawing such as center-lines and so on.

If you want, make a hard copy of the pulley drawing. If you have zoomed in or out on the drawing, make sure that the plot setup is set to *Full Plot* before creating the plot file.

Don't forget to save your drawing.

Conclusion

As you can see, although Pro/E handles most of the work in creating the geometry of the drawing, there is still a lot to be done manually regarding the esthetics of the drawing. It is for this reason that you need to be quite familiar with drawing practices and standards. Pro/E gives you a lot of tools for manipulating the drawing - we have only scratched the surface here. There is actually an entire volume of Pro/E documentation (several hundred pages) devoted expressly to creating drawings! All this information is available on-line. Some additional drawing tools and techniques are discussed in the *Pro/ENGINEER Advanced Tutorial* from SDC.

The most important lesson here is that the engineering drawing is a by-product of the 3D solid model. We don't so much "make a drawing" as much as just "show the existing model". We observed how bidirectional associativity works in Pro/ENGINEER. It is this capability that gives Pro/E and all its related modules so much power. If several people are working on a design, any changes done by, for example, the person doing the part modeling, are automatically reflected in the drawings managed by the drafting office. As you can imagine, this means that in a large company, model and file management (and control) becomes a big issue. Pro/E contains a number of other drawing utilities to make that management easier, but we will not go into them here.

A second lesson is that the dimensions that will automatically show up in the drawing are those used (for example, in Sketcher) to create the features of the model. Therefore, when creating features, you must think ahead to what information you want to show in the final drawing (and how). This involves your identification and understanding of the design intent of the features in the part. A part kludged together from disorganized features will be very difficult to present in an acceptable drawing.

We will return to creation of drawings in Lesson 10. There we will see some more tools and techniques to expand on the ones covered here.

In the next lesson, we will see how to create an assembly using the L-bracket and pulley you created in this lesson. We will also have to create a few small parts (washers, shaft, base plate).

Questions for Review

1.   When creating a new drawing, how do you specify which part is going to be drawn?
2.   Is it possible to create a 2D drawing without a part? What advantages/disadvantages might this have?
3.   The first view added to the drawing is called the _____?
4.   How do you set the orientation of a view? Consider both the first view and subsequent views.
5.   How do the mouse buttons function for dynamic view control in a drawing?
6.   What is the meaning of the following toolbar icons?

     a)   b)   [icon]   c)   [icon]   d)   [icon]   e)   [icon]

7.   What is the easiest way to move a view on the drawing sheet?
8.   Is it possible to delete a view once it has been created?
9.   On a very complex part, do you think that *Show All* is very useful? Why?
10. How can you edit the text contained in a note?
11. How do you change a section hatch pattern to a standard material?
12. When you select *With Preview*, what color do dimension details first appear in?
13. Explain the functioning of the three mouse buttons as used to modify dimension cosmetics.
14. How do you select a drawing template? What does it create for you automatically?
15. Describe two methods to move a dimension from one view to another. When might you want to do this?
16. How do you create a text note (for example, to put in a title block)?
17. Can you move or delete views created automatically with a drawing template?
18. What happens if you change the value of a dimension in the drawing?
19. What is the difference between *erasing* and *deleting* a dimension?
20. Is it possible to add new features, or redefine existing features when you are in drawing mode?
21. How can you produce hard copy of a drawing?
22. When you want to create a section view, at what point in the command sequence for adding the view do you designate it to be a section view?
23. What four items of information are required in order for Pro/E to generate a section view?
24. How can you turn off hidden lines in a section view?
25. What boundary options are available for creating a detail view?
26. When showing dimensions, what are the available **TYPE** options?
27. How can you change the spacing and angle of a hatch pattern?
28. How is the design intent reflected in a drawing, and how does this relate back to the part?
29. Do you think it would be possible to have a completely automatic system for creating a fully dimensioned drawing?
30. What symbol is used in a note to tell Pro/E to display the value of a parameter?
31. What other parameters (other than the three we used) are built into the system?
32. How are drawing options set?
33. What is the difference between a *template* and a *format*?

Exercises

Here are some parts to practice producing detailed engineering drawings. You may have created the models for the first two at the end of a previous lesson. Do these again, keeping in mind what you want to show on the drawing - you will probably make the model differently this time!

Project

Here is a cut-away view of the final major part of the project. Some dimensions will have to be inferred - use a reasonable estimate. See the VRML model on the SDC web site if required. Some careful planning for this part will pay off in reduced modeling time.

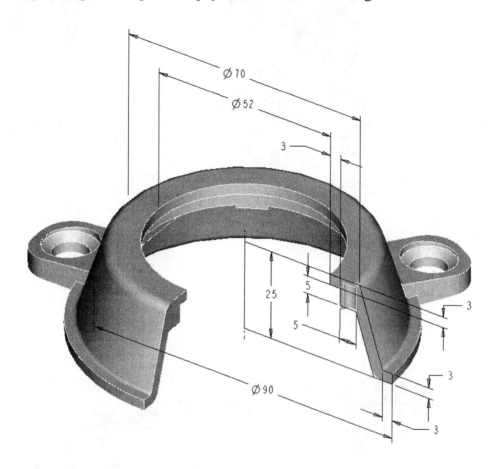

There are a couple more pictures on the next page.

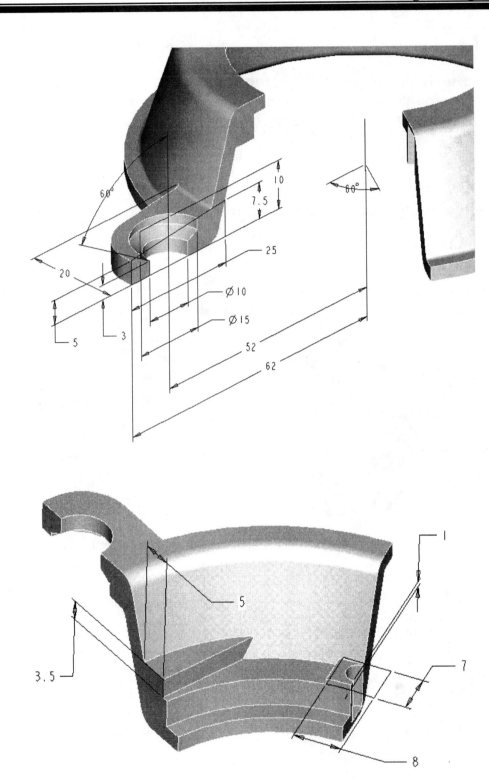

Lesson 9

Assembly Fundamentals

Synopsis

Introduction to assembly mode; assembly constraints; subassemblies; screen layout and assembly options; assigning colors to components.

Overview of this Lesson

In this and the next lesson, we are going to look at how you can use Pro/E to create and modify an assembly of parts. You have already created two of the parts involved: the pulley and the support bracket (see Lesson #8). In this lesson, we must first create a number of other simple parts needed for the assembly exercise - a couple of these will involve a new sketcher trick. Then we will use Pro/E to combine the component parts into an assembly. The finished assembly is shown in Figure 1. An exploded view of the components is shown in Figure 2.

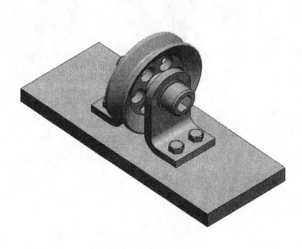

Figure 1 Final assembly containing 14 parts (some repeated)

Figure 2 Exploded view of final assembly

We will intentionally create some of the parts with dimensions different from those required in the final assembly so that in the next lesson we can go over some of the part/assembly modification commands..

The lesson is organized as follows:

1. Creating the Assembly Components
2. Discussion of Assembly Constraints
3. Assembly Design Issues
4. Assembling the Components
    ▸ Creating a Sub-Assembly
    ▸ Creating the Main Assembly
5. Assigning Colors

As usual, there are Questions for Review at the end of the lesson.

Creating the Assembly Components

IMPORTANT NOTE: Make sure all your parts have units set to millimeters.

The Pulley

As mentioned above, you should have created the pulley in Lesson #8. One thing we forgot to do then was add a keyway to the central hub of the pulley. Do that now: the keyway is *5mm* wide and about *3mm* deep. Create the keyway as a *both sides cut* off **FRONT**. Put the keyway at the 3:00 o'clock position (symmetric about **TOP**). The keyway should look like Figure 3.

You can leave the pulley in session when you move on to the next part. As a safety precaution, you should save it now.

Figure 3 Pulley with keyway added

The Axle

Create a part called **axle** as shown in the figure at the right. Use the dimensions shown in Figure 5 (we will change some of these later when we are in assembly mode). See the hint below for creating the hexagon.

Figure 4 The pulley axle

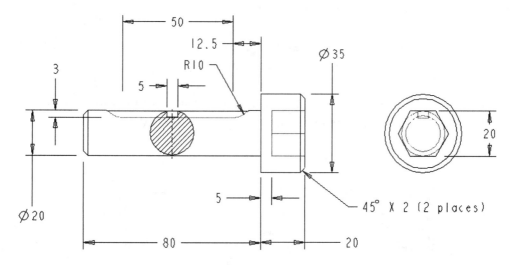

Figure 5 Dimensions for the axle

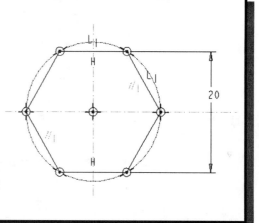

Helpful Hint

To make a hexagonal sketch a bit easier, first create a circle, highlight it (in red), then right click and select *Construction*. You can now sketch the six sides of the hexagon with vertices on the circle (observe the snaps that happen with Intent Manager). Add a dimension for the width across the flats. Then use the sketcher constraints to eliminate all the other dimensions. The final sketch is shown at the right.

The Base Plate

Create a part called **bplate** according to the dimensions shown. The plate thickness is *20* mm. Note the two planes of symmetry. How can you exploit this?

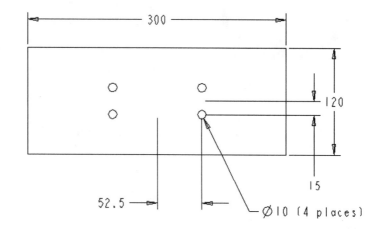

Figure 6 Base plate dimensions

The Bolts

We will need several bolts in the final assembly. These will all come from the same part file **bolt** containing only a single bolt. Note that the threads have not been included for simplicity here. If you wanted to include the thread, you could use a helical cut (making the screen display very slow) or using what is called a *cosmetic thread*. The dimensions of the bolt are shown in Figure 8. See the hint above for creating the hexagonal sketch for the head.

Figure 7 The bolt

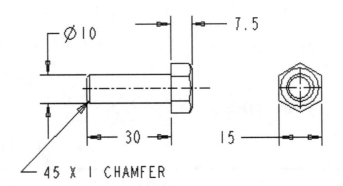

Figure 8 Bolt dimensions

The Bushings

We will need a couple of these too - call the part **bushing**. It is a simple protrusion.

Figure 9 The bushing

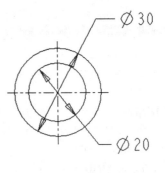

Figure 10 Bushing dimensions

The Washers

Our last component part is the washer. It has the dimensions
shown in the figure at the right. The easiest way to make this is to
do a *Save A Copy* of the bushing part (giving a new name
washer), *File > Open* the new part, then double-click on the
protrusion and edit the dimensions. You may want to change the
DESCRIPTION parameter as well:

> *Tools > Parameters*

Click on the **Value** entry for the parameter DESCRIPTION and
enter a new string.

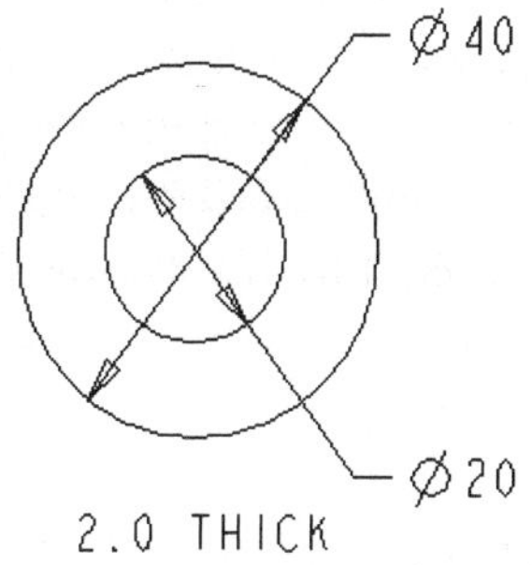

Figure 11 Washer
dimensions

When you start assembling these components later in this lesson, make sure they are all in your
start-up directory. And (just another reminder!) all these parts should be in millimeters.

Assembly Constraints

Creating an assembly is actually a lot of fun and not too difficult. Your main challenge will be
display management as the screen gets more cluttered with objects. Creating an assembly
involves telling Pro/E how the various components fit together. To do this, we specify *assembly
constraints*.

The geometric relation between any two parts has six degrees of freedom: 3 translational and 3
rotational. In order to completely define the position of one part relative to another, we must
constrain or provide values for all these degrees of freedom. Once we give Pro/E enough
information it will be able to tell us when the part is fully constrained and we can assemble the
part. We proceed through the assembly process by adding another part, and so on.

A component that is fully constrained in the assembly is called *placed* or *assembled*. It is
possible to leave a component not fully constrained, in which case it is called *packaged*. Pro/E
will be able to tell you whether a new component is packaged or assembled. Packaged
components will be easy to identify with a special symbol in the model tree. We normally avoid
leaving components in this condition. Components that have at least one constraint to a
packaged component are automatically considered as packaged only.

There are a number of constraint types that we can specify. In this lesson, we will use six of
them. The rest should be pretty easy to figure out on your own. The individual constraints are
used with the surfaces, axes, datum planes, and datum points of the components involved in the
assembly. The constraints usually must be used in combinations in order to fully constrain all 6
degrees of freedom. Here are the main constraint types that we will discuss:

MATE COINCIDENT (or just **MATE** for short)

Two planar surfaces or datums become coplanar and face in opposite directions. When using datums, you must specify which side is involved. This constrains 3 degrees of freedom (one translation and two rotations). Can you think what they are? There are still 3 unconstrained degrees of freedom (what are they?).

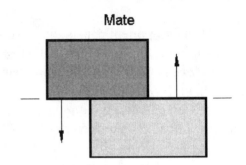

Mate

Figure 12 The *MATE* constraint

MATE OFFSET

Two planar surfaces or datums are made parallel, with a specified offset distance, and face in opposite directions. The offset dimension can be negative, and can be used in assembly relations to automatically change the distance between the surfaces. What degrees of freedom does this constraint fix? Which ones are still free?

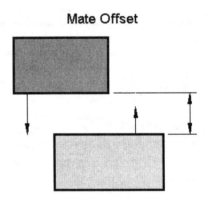

Mate Offset

Figure 13 The *MATE OFFSET* constraint

ALIGN COINCIDENT (or just **ALIGN** for short)

This can be applied to planar surfaces datums, revolved surfaces and axes. Planar surfaces become coplanar and face in the same direction. How many degrees of freedom does this constrain? When aligning datum planes, you will have to specify which side is to be aligned.

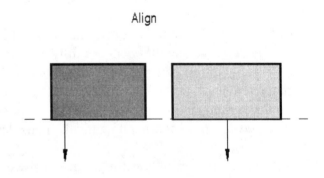

Align

Figure 14 The *ALIGN* constraint with planar surfaces

When Align is used on revolved surfaces or axes, they become coaxial. How many degrees of freedom are constrained? Also, note that there are still two possible positions (obtained by reversing the direction of one of the axes) - you can force one or the other with the Orient constraint described below.

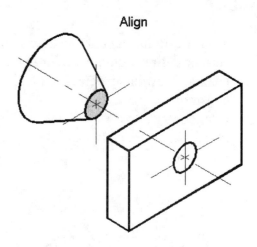

Figure 15 *ALIGN* used with surfaces of revolution aligns the axes

ALIGN OFFSET

This can be used only with planar surfaces: they become parallel with a specified offset and face the same direction.

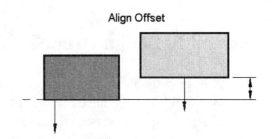

Figure 16 *ALIGN OFFSET* (used with planar surfaces only)

ALIGN ORIENT

Two planar surfaces or datums are made parallel and face the same direction (similar to Align Offset except without the specified offset distance). How many degrees of freedom does this constrain?

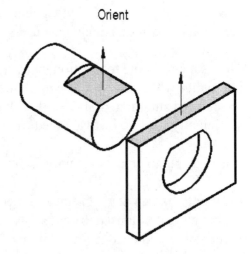

Figure 17 The *ALIGN ORIENT* constraint

INSERT

This constraint can only be used with two surfaces of revolution in order to make them coaxial. How many degrees of freedom does this constrain?

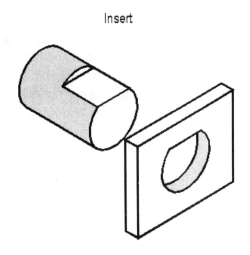

Insert

Figure 18 The *INSERT* constraint used with cylindrical surfaces

Assembly Design Issues

Before beginning an assembly (or even when you are creating the parts), you should think about how you will be using these constraints to construct the assembly. As we saw in Lesson #1, the logical structure of the assembly is reflected in the model tree. Like designing the features of a part, the chosen assembly constraints and references should reflect the design intent. It is possible to create an assembly that fits together, but if the chosen constraints and references do not match the design intent, changes that may be required later could become very difficult. Pro/E does provide tools for dealing with this (the 3 R's also work in assemblies on the assembly references of components), but you should really try to think it through and do it right the first time! Obviously, it takes a lot of practice to do this. The "I-should-have-done-it-*that*-way!" is a common realization among new users.

This is a good time to mention that when you are placing a component into an assembly, it does not matter what order you use to define the placement constraints for the component, since they are applied simultaneously. Pro/E will tell you when you have constrained the component sufficiently for it to be assembled. The order you use can be chosen strictly for convenience.

It is possible to create assembly features (like datum planes and axes, and even make datums) that will exist only in the assembly. This would allow you, for example, to use an assembly parameter like an angle or linear dimension between datums, to control the assembly geometry. In this way, if you used the assembly feature as a constraint reference for a number of component parts, you could change the position of all parts simultaneously in the assembly by modifying that parameter. All the parts would still have to assemble according to all the assembly constraints defined between parts.

In the context of this lesson, an assembly consists of a number of components that are rigidly constrained to each other - no moving parts! There are additional functions and software modules in Pro/E that allow you to create assemblies of moving parts, that is, mechanisms. This extension lets you create different types of joints and connections (pin, slider, ball, cam, and so on) between components, and drivers to control the degrees of freedom. You can then analyze the motion of components and even create animations of the moving system.

Finally, you should note that Pro/E will happily let you assemble two components that interfere with each other (have partially overlapping solid volumes). Although the parts are "solids", Pro/E does not prevent you from assembling them with interference. Sometimes, in fact, this is the desired result (as in a shrink fit, or designing built-in snaps and catches in plastic parts). More usually, interference results when dimensional values are not correct (or finalized) in a design. Pro/E has a simple tool that will tell you if any components are interfering (and by how much). It is usually a simple matter to correct the dimensions to remove interference once the assembly is put together.

Assembling the Components

Before you begin, make sure that the parts **lbrack.prt** and **pulley.prt** that you made in Lesson #8 are available in the working directory along with the parts you made earlier in this lesson.

Creating a Subassembly

We will start by assembling the L-bracket, a bushing, and a washer into a subassembly. This will save us some time, since two copies of this subassembly must be inserted into the final assembly. Once created, a subassembly is treated exactly the same way (in regards to subsequent placement constraints) as a single part.

From the **FILE** menu, select

> *New > Assembly | Design > [support]*

Deselect the **Use default template** option and select **OK**. In the **New File Options** window, choose the *Empty* template and again select **OK**. Close the model tree (it's empty now anyway).

In the pull-down menu, select

> *Insert > Component > Assemble*

or pick the *Add Component* button in the right toolbar. Either way, select the part **lbrack.prt** and then **Open**.

Since this is the first component we are bringing into this assembly, there is nothing to constrain it to. Pro/E therefore puts it in the assembly in the same default orientation as the component itself. This is all right for this part, but may not always be. To have more control over the orientation of the first component, we would like to be able to constrain it. We will do that for the next assembly we make.

You can turn off the datum planes if you like. Now we'll add the bushing. Select the *Add Component* button again and *Open* the part **bushing.prt**.

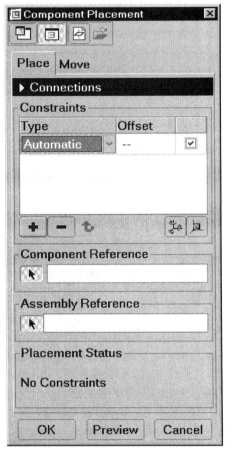

The bushing will appear somewhere beside the bracket and the **Component Placement** window (Figure 19) will open up. This window will list the various placement constraints as they are created for this component, and allow us to select constraint types and references on the new component and the existing assembly. Note that the component (ie. the bushing) is not fully constrained - see the **Placement Status** at the bottom of the window.

There are two main display modes when you are doing assembly. We will look at both of them. These modes are set by the two buttons at the top left of the new window. We'll call these the **Separate Window** and **Same Window** buttons. First, make sure that only the *Separate Window* option is selected. We will use the other display mode button a bit later. This puts the current assembly in one graphics window (title: *SUPPORT*), and the component being added in another (title: *BUSHING*). Resize the two windows so that they do not overlap. Having both windows open makes it easy to locate references and gives us independent viewing control over the zoom/spin/pan in the two windows. This is useful, for example, when dealing with a small component in a large assembly structure.

Figure 19 The **Component Placement** window

Read the following few paragraphs before proceeding:

Constraining the component involves three steps:

1.   In the **Constraints** area of the Component Placement window, you can select the desired **Constraint Type** from the pull-down list. Selecting *Automatic* (the default) allows Pro/E to determine the type of constraint you probably want based on the type of entity you choose. For example, picking two axes basically means you want to do an *Align* (since *Mate* makes little sense with axes). You can, of course, override this constraint. If you pick two surfaces, the automatic choice may be an Align, but you can easily change this to a Mate by re-selecting the constraint in the pull down list.
2.   In the *component window*, you will select the appropriate reference surface, datum, or axis for the constraint.
3.   In the *assembly window*, you will select the matching reference surface, datum, or axis for the constraint.

As you add constraints, keep your eye on the **Constraints** list box, and the **Placement Status** area. At the right of the Constraints list is a check box so that you can disable/enable individual constraints after they have been created. In previous releases of Pro/E, it was not possible to overconstrain a component. That is now possible (as long as the constraints are consistent). You will be told when you have provided enough constraints for the new component to be fixed in the assembly. You have to be a bit careful here, since it will sometimes be possible to include the component at what appears to be the correct position without it being entirely constrained. The **Component Placement** window will let you exit without fully constraining the component; as mentioned earlier this is called "packaging" the component. Unless you really want to do this, make sure the **Fully Constrained** status appears before leaving this window. Also, remember that the order of creating the constraints does not matter, nor does the order of picking references on the component or assembly.

If it is difficult to see or select an entity to be used for an assembly constraint in the model's present orientation, either in the **Assembly** window or in the **Component** window, you can use the mouse buttons to spin/zoom/pan the part or assembly as usual, after clicking in the appropriate graphics window. This can be done at any time while specifying a pair of placement constraints. Preselection highlighting also comes in handy here.

For the bushing, we want to set the constraints shown in Figure 20. The *Insert* constraint makes the cylindrical outer surface of the bushing line up with the surface of the hole[1]; the *Align* constraint keeps the face of the bushing even with the surface of the bracket. Before proceeding with applying these constraints, resize/reorient/move the bracket and bushing displays so that you will be able to easily pick on the appropriate entities.

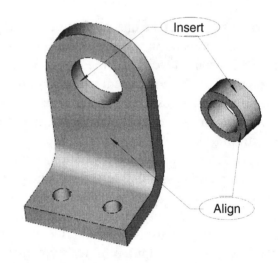

Now we'll proceed with the assembly. Select (in the pull-down list)

> #### Constraint Type (Insert)

Figure 20 Constraints for the bushing

and pick on the outer surface of the bushing. It highlights in red. Read the bottom line in the message window of the assembly. Pick on the inner surface of the large hole in the bracket. In the **Component Placement** window, you should see a new line entry for the constraint, and the message that the component status is "Partially constrained". The bushing can still slide along, and rotate around, its axis. Now select

> #### Constraint Type (Align)

and pick on the flat face of the bushing. Then pick on the flat surface of the bracket. The message

---

[1] As mentioned before, *Insert* actually causes the axes to align. The two surfaces we select for *Insert* do not have to be physically in contact with each other.

in the **Component Placement** window will inform you that the component is fully constrained. Note that the "Allow Assumptions" box (which came out of nowhere!) is checked. What does this do? Is the bushing, in fact, fully constrained at this time? The answer is no (!), since the bushing is still free to rotate around its axis. Pro/E has determined, with an assumption, that this degree of freedom doesn't matter for this part. Deselect the **Allow Assumptions** box. Now Pro/E tells you that, indeed, the bushing is not fully constrained. What would be required to complete the constraints? (Hint: turn the datums back on.) Don't add this constraint now, since the assumption isn't going to hurt us. Turn the **Allow Assumptions** box back on.

You can then select *Preview* to see where Pro/E will put the component in the assembly. The placement will be indicated in yellow wireframe as shown in Figure 21. The bushing should be even with one side of the bracket and protrude slightly from the other (since it is a different thickness than the bracket).

When all constraints are complete, click on different constraints listed in the **Constraints** box. The various surfaces involved in each constraint will be highlighted.

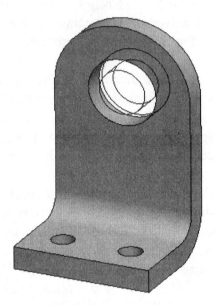

If you make a mistake in specifying the type or references of a placement constraint, you can select it in the **Constraints** box. For example, click on the *Insert* entry. The associated references on the component and assembly are shown in purple and blue, respectively. These are easiest to see in wireframe or hidden line display. Then, either *Remove* the constraint (use the button with the red "-" sign), select a new type, or select the small arrow buttons beside the listed component and assembly references and pick new ones.

Figure 21 Bushing assembled to the L-bracket

When you want to create a new constraint, make sure to first select the *Add* button (use the button with the green "+" sign), otherwise you may end up redefining an existing constraint. This button is automatic unless you have interrupted the normal flow of assembly steps.

Observe the effect of the checked boxes on the right of each listed constraint.

If you are happy with the bushing placement, select *OK*. The graphics window will now show the L-bracket with the bushing in place.

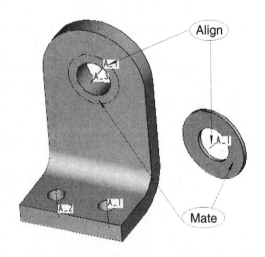

Now we will place a washer on the outside of the bushing. Select the *Add Component* button and pick the **washer.prt**, then *Open*.

Figure 22 Placement constraints for the washer

This should again open in a separate window.

We will create the placement constraints shown in the figure at the right. These constraints are *Align* (washer and bushing axes) and *Mate* (washer and bushing faces). Both of these are *Coincident* by default.

First - the axes. How can you make sure you are picking the axis of the bushing and not the hole in the bracket? In the assembly window, these coincide. Set the **Selection Filter** to *Axes*, then right click on the bushing/hole axes - one of them will highlight. Now hold down the RMB and select *Pick From List*. This opens a window that lists both datum axes at the pick point. Select the axis for the bushing (you might have to scroll right in the list to confirm this), and then *OK*. In the **Component Placement** window, the assembly reference should indicate the bushing part.

For the Mate constraint, pick on the flat surface of the washer, then on the end surface of the bushing. Make sure the constraint type is set to *Mate*. You may have to *Flip* the constraint.

Leave **Allow Assumptions** turned on and *Preview* the assembly. Select *OK* when you are satisfied.

We are finished creating this subassembly, so select

> *File > Save*

and select the default name for the assembly (**support.asm**).

AN IMPORTANT NOTE

We observed above that the default orientation of the support assembly is the same as that of the first component (the L-bracket) brought into the assembly. This is because there was nothing to constrain the bracket to. We are stuck with a default assembly orientation that was basically defined at a part level. It is seldom a good thing to limit our options this way. Furthermore, none of the saved views or other common parameters used in assemblies are defined. We will address both these concerns in the next assembly we make by using an assembly template. This will provide us with datum planes to constrain the first part to (in whatever orientation we want) plus all the standard views (TOP, FRONT, RIGHT, etc.).

Creating the Main Assembly

Leave the subassembly window open, and create a new assembly called **less9** using an assembly template as follows

> *File > New > Assembly | Design >[less9]*

Deselect "Use default template" and select *OK*. Pick the assembly template *mmns_asm_design*. Enter values for the parameters DESCRIPTION and MODELED_BY and select *OK*. This has default datums (notice the names), a couple of parameters, and the usual predefined views. Unless you have a really good reason not to, you should always use a template or at least create the default datums. The advantage of the template is the saved views plus it gives us complete freedom in defining the orientation of the first component we define in the assembly.

Bring in the first component, the base plate. Use the ***Add Component*** button in the right toolbar and select the part ***bplate.prt***. Keep using the **Separate Window** option for now. We need to constrain this component to the assembly datums. We will experiment with that a bit, showing three ways of doing it. Each will result in the same position/orientation of the part in the assembly.

First, leave the constraint type set to **Automatic**. Pick any of the datums in the component then pick on the corresponding datum in the assembly (component **RIGHT** and assembly **ASM_RIGHT**, for example). Pro/E automatically sets up an ***Align Coincident*** constraint. Do this for each pair of datums. The component should now be fully constrained. We could, of course, have created any correspondence with the three datums, as long as they were consistent, to reorient the base plate however we like.

To see a second method of constraining the plate, delete these three constraints (with the "-" button), then select the ***Add*** button "+". The constraint type is ***Automatic*** again. Set the **Selection Filter** to ***Coordinate System***. Pick on the coordinate systems in the component and the assembly. This aligns the XYZ axes of the two coordinate systems. Pretty easy!

For a third option, delete the ***Coord Sys*** constraint (with the "-" button), and select the ***Add*** button "+" again. If you actually want the default placement and orientation of the assembly to be the same as the component, select the ***Assemble Default*** button [⊒] .

You can accept the component placement with ***OK***. We don't need the datum planes or coordinate systems any more, so you can turn off their display.

Now bring in the subassembly. We add this just as if it was a single component. Select the ***Add Component*** button and pick the ***support.asm*** sub-assembly in the directory listing. We will set up the placement constraints for the subassembly shown in the Figure 23.

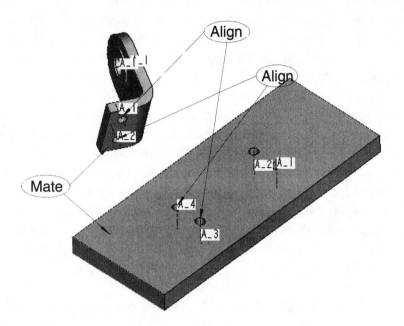

Figure 23 Placement constraints for the subassembly

First, pick the lower surface of the bracket with the upper surface of the base plate. The default constraint for two solid surfaces is *Mate Coincident*. That is fine here. Then select the axis of one of the bolt holes in the bracket with the axis of the appropriate hole in the base plate. The default constraint here is *Align Coincident*. If **Allow Assumptions** is turned on, you will get the message that the component is fully constrained, so select *Preview*. You will see the support shown something like Figure 24 (this will depend on which holes you chose to align, and on how you oriented your parts when you created them).

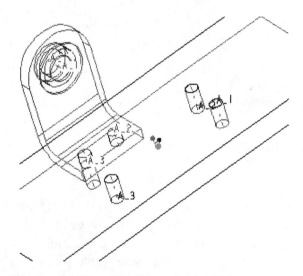

Figure 24 First MATE and ALIGN constraints applied to subassembly

Hmmm... not exactly what we want. The bracket is not actually fully constrained yet, since it can still rotate around the hole. Turn off the **Allow Assumptions** option. Select *Add* ("+") in the **Component Placement** window) and pick the other bolt hole axis in the bracket and the appropriate axis on the base. This will appear as *Align Oriented*. The sub-assembly is now fully constrained without any assumptions. Select *OK.* You can minimize the *support.asm* window now and resize the window containing the assembly.

We'll now bring in another copy of the support subassembly using *Assemble*, and attach it to the base plate so that it faces the first one as shown in Figure 28. We will use a slightly different screen display and options. Select the *Add Component* button and once again pick the component *support.asm*[2].

In the **Component Placement** menu make sure that only the **Same Window** button is selected. The second subassembly will appear in the same window as the base plate. We are going to align the axes and mate the surfaces as before using the other set of holes in the base plate. This time, however, the display will show the position of the subassembly relative to the assembly as each new constraint is added. This is a convenient way to keep track of the effect of your assembly constraints, however since everything is happening in one window it may be difficult sometimes to select references. *Preselection Highlighting* and making effective use of the Selection Filter and Selection List are indispensable here. Your assembly sequence might look like Figures 25 through 28.

For the first constraint, select the axes of the holes that will end up at the front of the plate.

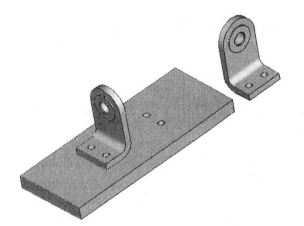

Figure 25 Subassembly brought into session

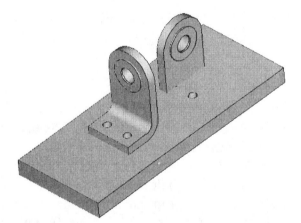

Figure 26 First hole axes aligned - support buried in base plate!

Notice that in Figure 26 there is overlap (interference) of the bracket and base plate. After you get to this stage, observe the effect of the different view options (wireframe, hidden line, no hidden, shaded) as you spin the assembly. The displays may not be what you expect. You should be able to recognize these view effects as symptoms that you have interfering components[3].

---

[2] A good question at this point is: Why not just mirror the existing support? When you mirror a component in an assembly, you create a new component. In this case, it would make a new sub-assembly. It would also make mirror copies of all components in the sub-assembly. This is actually a valuable tool to know if that is really what you want. In this case we don't - we want the same component used twice. This will keep the necessary number of components and files to a minimum, and will result in the appropriate entries in the assembly Bill of Materials.

[3] Pro/E has other much more sophisticated tools for detecting interference.

When you reach the configuration shown in Figure 26, hold down the CTRL and ALT keys simultaneously while you pick the new support. This lets you reposition the support (by dragging it with the mouse) while maintaining the existing constraints. Drag it above the base plate where you can easily pick its lower surface. Then pick on the top surface of the base plate. The *Mate* constraint will occur automatically. See Figure 27.

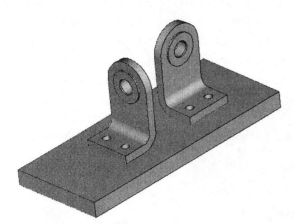

Figure 27 Surfaces mated. Fully constrained with assumptions. No overlap.

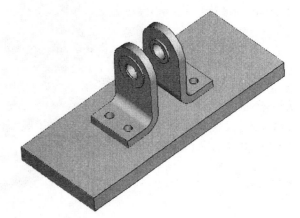

Figure 28 Second bolt axis aligned with other hole on base plate

Finally, pick on the second hole axes on the bracket and the base plate. This should get you to the configuration shown in Figure 28. Save the assembly.

Helpful Hint

When you are using *Same Window*, holding down the CTRL and ALT keys while you pick on the component being assembled lets you drag it around the assembly. Its movement is restricted by whatever constraints have been defined already.

Now we'll assemble the axle using the following constraints. Read ahead through the next couple of pages, since we are going to do something a bit different again.

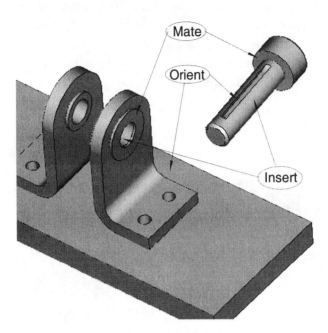

Figure 29 Placement constraints for axle

Before we do the assembly, let's review our constraint design. The *Mate* constraint is between the bottom of the axle head and the outer face of the washer. The *Insert* constraint could be with any of the inner surfaces of the bushings or washers on either support. The design intent will be best served if you pick a surface of a bushing. In either case, this constraint will allow the component to be placed, but it will still be able to rotate around its own axis. We'll add another constraint to prevent this by *Orient*ing the lower surface in the keyway and the upper surface of the base plate.

Bring in the axle and make sure the **Same Window** button is selected for the component display. The axle will appear somewhere, perhaps similar to Figure 30.

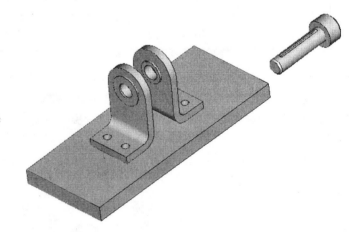

To control the display while placing the axle, a useful tool is available to rearrange components on the screen. Select (the tab at the top of the **Component Display** window)

Move
Translate | View Plane   **Figure 30** Axle selected for assembly

Read the message in the command window. Click on the axle and drag it to a position similar to the one shown in Figure 31.

Now, for **Motion Type** select the *Rotate* option, in the **Motion Reference** list select *Entity/Edge*, click on the axis of the axle, and spin the axle by dragging with the left mouse button. You should be able to spin it a full 360°. See Figure 32. Rotate it again and drop it in the initial position with the keyway on the top of the axle.

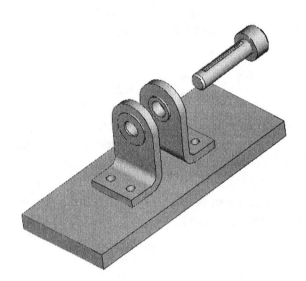

Figure 31 Axle after translating from initial placement position

Figure 32 Axle after rotating around its axis

At the same time as you are moving the component, you can control your view (spin, zoom, and pan) using the dynamic view controls as usual. This gives you considerable control over what you see on the screen. You can also translate and rotate relative to surfaces and axes in the assembly.

You will need to experiment with these controls for a while before you will be comfortable with them.

When your display shows you a convenient view of the axle and the assembly together, select the *Place* tab and set up the assembly constraints indicated above. We are going to apply the constraints in the order: *Insert*, *Align Orient*, *Mate*. As mentioned earlier, the order of creating these constraints doesn't matter to the final placement. You will find some sequences easier to implement than others. For example, try to avoid the "buried" phenomenon we encountered earlier (for the support subassembly) that makes it hard to select references. The *Insert* constraint should not be any trouble. Set that up first.

As you apply the constraints, try to *Move > Translate* and *Move > Rotate* the part. Also recall the hint about using CTRL+ALT. You will find that these moves are restricted because of the existing constraints at the time.

For the second constraint (orientation of the keyway), in the **Constraint Type** list, start a new constraint by over-riding the *Automatic* setting and selecting *Align*. Now click on the box to the right, in the **Offset** column. In the new pull-down list that appears, select *Oriented*. Now pick the surface at the bottom of the keyway on the axle. Then pick the top surface of the plate. The axle will rotate to the desired position.

Finally, pick on the two surfaces for the *Mate* constraint. We have saved this one for last so that the surfaces are clearly visible and easy to pick out. If you are asked for an offset, enter *0*. This enters a parameter (the offset value) into the model that you can *Edit* later. Otherwise, in the **Offset** column of the *Mate* constraint, open the pull-down list and select *Coincident*. This removes the parameter.

VERY IMPORTANT NOTE:
> The *Move* command (previously called a "Package Move") is used for cosmetic purposes only. Although it may be possible to move a new component into the correct position relative to other parts, you must still specify the geometric constraints in order to assemble it. If you leave the **Component Placement** window without fully constraining the component, it is called "packaged." A special notation will appear in the model tree for such a component. **A new component that is constrained (even partly) to a previously packaged component will itself be considered packaged only.**

The final position of the axle should be as shown in Figure 33. Notice the position of the keyway.

If everything is satisfactory, select *OK*. Otherwise, click on a constraint listed in the table, select either the constraint type, component reference, or assembly reference, and make the appropriate corrections.

Now is a good time to save the assembly.

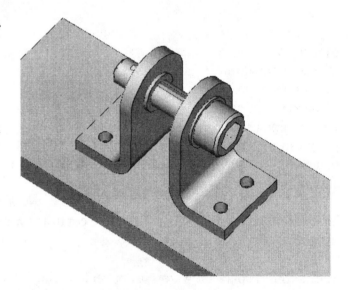

Figure 33 Final placement of axle

We can now bring in the pulley and attach it using the constraints shown in Figure 34. You might like to experiment with *Separate* and *Same* window displays, and possibly use shaded views to help identify surfaces. This is useful when the assembly starts to get crowded with visible and hidden edges, datum planes and axes, and so on.

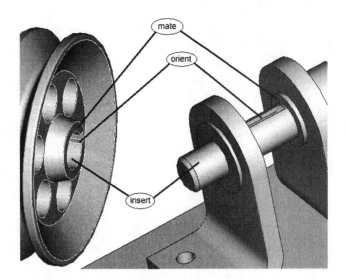

Figure 34 Placement constraints for the pulley

Once again, the pulley could be placed with just the *Insert* and *Mate* constraints. But, we want to make sure the keyway lines up with the axle. The *Align Orient* constraint can be used with a side surface of the keyway, and a side surface of the keyway on the axle. When the pulley is assembled, it should look like Figure 35.

Sometimes, when you pick an axis or surface alignment, Pro/E decides to place the component 180° from where you want it. In that case, select the constraint and try the *Flip* button ⬆️.

Figure 35 Final position of pulley

Finally, bring in the four bolts to attach the brackets to the base plate. We'll bring these in one at a time for now - there are a number of advanced assembly commands that would allow you to create a pattern of bolts that would match a pattern of bolt holes. This would allow the assembly to automatically adjust, for example, if the pattern of bolt holes in the base plate was changed (including changing the number of bolts in the assembly). To place a single bolt, the placement constraints are shown in Figure 36. You can *Allow Assumptions* for these bolts.

Experiment with the *Separate Window* and *Same Window* options, and try out the *Move* commands some more. You have four bolts to experiment with. Place a bolt in each of the holes available. The assembly is now completed and should look like Figure 37.

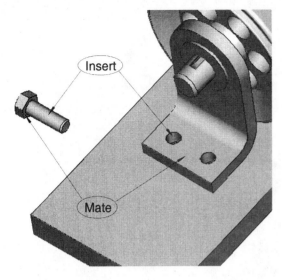

Figure 36 Placement constraints for bolts

Figure 37 Completed assembly

You should continue to experiment with Preselection, Filter Settings, the Move commands, and display modes. Most users prefer the **Same Window** mode of operation for specifying constraints, however the **Separate Window** is very useful if you are having trouble picking references in a part, or if you are assembling a very small part to a very large assembly.

Save the assembly. Open the model tree and explore the information presented there.

Assigning Appearances to Components

When a component or assembly is displayed in shaded mode, or when a rendered image is created, its appearance is determined by several settings we apply. The most important of the appearance settings is color. Other aspects of an appearance definition are texture, transparency, reflection, shadows, and so on. Appearances can be applied to entire objects or individual surfaces. Defined appearances can be stored in files. A default appearance definition file (*appearance.dmt*) is loaded at program start-up. Once an appearance has been assigned to a component, its definition is contained with the component file. This means if the component is sent to another system the component's appearance will be taken with it.

In the following, we will deal only with color. Make sure that colors are turned on (*Tools > Environment*).

We assign appearances (colors) in two steps: first we define the colors we are going to use, then we apply the colors to the desired components. The extent to which you can do this will depend on the specifics of your Pro/E installation and your hardware. Select:

> *View > Color and Appearance*

The **Appearance Editor** window will open as shown in Figure
38. The pane below the pull-down menus shows the appearances
currently defined on your system. This is often called the *color
pallette*, but as was mentioned above each appearance contains
much more than just color. By default, these appearances are
stored in a file called *appearance.dmt* in the default working
directory which is loaded when Pro/E is launched. If it is missing,
only one color - white - is defined. Each appearance will have a
name. The name of the default appearance in Figure 38 is
ref_color1. If you have loaded another assembly that has defined
appearances, they will be displayed here. If you have an
appearance file stored somewhere else, this can be loaded with
File > Open.

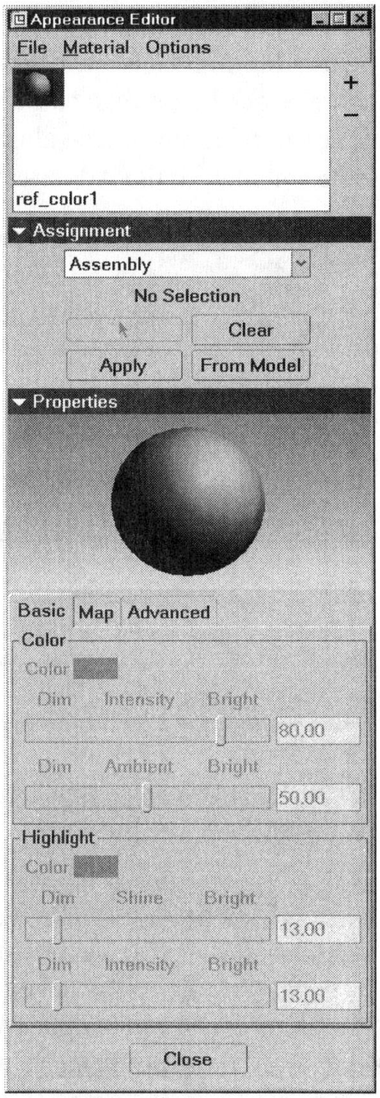

Investigate the pull-down menus to see what commands are there.
Some of the options may be hardware dependent. Just below the
color pallette area is the **Assignment** area. This is where we pick
what we want to assign appearances to. This can be an entire
assembly, separate components, or individual surfaces. Below this
is the **Properties** area, which is where we can define or change
the definition of an appearance selected in the pallette.

Let's define some more colors. Beside the color pallette, select the
Add button "+". The **Properties** area is now live. In the **Basic**
sheet **Color** area, the two sliders control the intensity of the
lighting on the model. Click the color patch button. This opens a
new window called the **Color Editor**. This editor contains three
different ways to select a color. The default method is to select
values of red-green-blue (RGB) (ranging from 0 to 255) in the
new color. You move the sliders until you get the right mix of
RGB for the new color, or enter integer values in the range 0 - 255

Figure 38 The **Appearance
Editor** window

in the boxes on the right. As you use the sliders to change the
value of any color component, the new color is shown at the top of the **Color Editor**, and on the
shaded ball back in the **Appearance Editor**. Another method of setting color is to use the **Color
Wheel**. This shows a disk containing all the available colors. The current color setting is shown
by a small cursor. You can click within the color wheel to pick the new color - this automatically
sets RGB values for you. Note that there is also an HSV (hue - saturation - value) method of
selecting color. If you select this method with the color wheel open, you can see the effect of the
three HSV sliders. The Hue slider moves the cursor in circles on the color wheel. The Saturation
slider moves the cursor radially on the color wheel. The Value slider sets the range between
black and the color on the color wheel.

Once you have selected a color in the **Color Editor**, select *Close*. The new color will show up in
the palette at the top of the **Appearance Editor** window where you can type in a name for the
new appearance. Click the *Add* button several times. This gives you a number of copies of the
currently selected color. You can pick each one individually and redefine its RGB definition,
rename it (make sure you hit the *Enter* key after renaming), and so on.

Define the appearance colors in the table below (these are the "primary" colors) or make up your own color mix. Color selection is a matter of personal preference. However, in general, lighter colors work better than darker ones. Appearance names might correspond, for example, with different materials (steel, aluminum, plastic, ...) If you have appropriate hardware, you might experiment with the **Advanced** menu to set transparency and other appearance attributes.

Color	RGB Composition		
	Red	Green	Blue
Red	255	0	0
Green	0	255	0
Blue	0	0	255
Yellow	255	255	0
Cyan	0	255	255
Magenta	255	0	255

When you have created the colors, *Close* the Color Editor. Back in the **Appearances** window, you can use

File > Save As

to save your newly created appearances. Remember that if you want this loaded automatically, it's file must be called *appearance.dmt* and be located in the start-up directory. You can use *File > Open* in the **Appearances** window to append colors in a stored file into the current session. There may be a system restriction on the total number of colors you can define - see your system administrator for details.

To apply color to the axle, select a color in the palette, then in the **Assignment** pull-down list, select **Components**. Pick on the axle then middle click and select *Apply*. If the axle color doesn't change, make sure that *Colors* is checked under *Tools > Environment*.

Choose different colors and assign them to the pulley, the base plate, and the four bolts.

When the total assembly is active, we can't individually color the components in the subassembly **support.asm** - if we tried that now, they would all end up the same color since this is treated as a single component in the current assembly. We will have to have the subassembly in its own window. If it currently isn't in your session (if it is you can do this by *Window > support.asm*), bring it in with

File > Open > support.asm

or if it is already loaded, just click on the window containing the subassembly and activate it.

Now you can set the colors of the constituent components. Once you have set all the colors, save the **support** subassembly, and change back to the overall assembly window. If you previously colored the support in the main assembly, you will have to *Unset* that color. Colors assigned at the highest level in the assembly tree take precedence. In this regard, you should note that colors can be defined and assigned at the individual part level. These colors are carried with the part into the assembly where they will stay unless over-ridden. For multiple occurrences of a part (like the bolts), it is easier to assign colors at part level, where you only have to do it once!

See how the display changes for wireframe, hidden line, and shaded displays. In wireframe display the edges of each part are shown in the assigned color. This might be awkward if you want to do any editing of the part, since line color is so important in representing information like highlighted edges, constraint surfaces, parent/child relations, and the like. To turn off the color display, select

Tools > Environment > Colors | OK

All edges will now be shown in the default colors. Turn colors back on again. You may find yourself toggling the color display state often enough that you might like to add a toolbar button to do that. See the Appendix on customizing the toolbars.

We are finished with the first lesson on assemblies. Don't forget to save your assembly - we'll need it in the next lesson. An important thing to note is that when you save the assembly, any component that has been changed will also be saved automatically.

You will note that the keyway in the axle extends beyond one of the support bushings. Also, the base plate is quite large. In the next lesson we will see how to modify an assembly and its component parts. This will involve creating assembly features (ie. specific to the assembly), as well as making changes to the parts themselves. It is also possible to create new parts while you are in assembly mode (we'll make the key this way, to make sure it fits in the assembly). We'll also find out how to get an exploded view of the assembly, and set up an assembly drawing.

Questions for Review

1.   If several identical parts are required in an assembly, do you need a separate part file for each one?
2.   What are the main assembly constraints?
3.   What is a "packaged" component? How does this restrict what you can do?
4.   What degrees of freedom are constrained by each of the main assembly constraints? Draw a sketch and illustrate the constrained and unconstrained degrees of freedom.
5.   What entities can be used when specifying assembly constraints?
6.   What is the difference between applying assembly constraints to individual components versus a subassembly?
7.   What is the difference between *Separate Window* and *Same Window*? Where are these options located?
8.   How do you select the constraint types?
9.   Can you do the assembly operations with a shaded view?
10. If you are in the process of constraining a component and you make a mistake, how can you a) delete, or b) edit a constraint.
11. What does *Move* do, and how is it related to applying assembly constraints?
12. Does it matter what order you create assembly constraints? When you are picking references does it matter if you select component references first?
13. Find out how many colors you can define on your local system.
14. Which takes precedence: colors assigned at part level or at assembly level?
15. How can you specify the colors of individual components in a subassembly?
16. How do you turn off color display in wireframe mode?
17. Which icons disappear from the right toolbar when you open or create an assembly?
18. What aspects of the display are controlled by "appearances"?
19. What (and where) is the default appearance file? What does it do? How can you modify it?
20. How do you change the color of an object?
21. Which takes precedence - an appearance assigned at the part level or at the assembly level?

Project

Start assembling the vise with this subassembly. Think about an assembly strategy and which constraints you are going to use before you starting actually doing anything. Look ahead to the finished product to see how you will be able to constrain this subassembly into the vise.

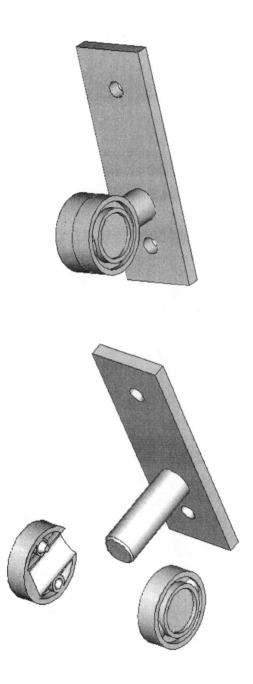

This page left blank.

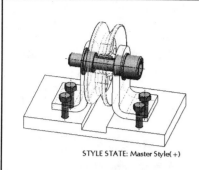

Lesson 10

Assembly Operations

STYLE STATE: Master Style(+)

Synopsis

Examining the assembly database. Declaring active components. Modifying parts in an assembly. Creating parts in assembly mode. Assembly features. Exploded views and display styles. Creating sections. Assembly drawings.

Overview of this Lesson

In this lesson, we will continue to work with the pulley assembly we created in Lesson #9. We start by using some Pro/E utilities to get information about the assembly (model tree, assembly references, assembly sequence). We will then see how to add features to the assembly, modify the parts by changing dimensions and adding features, and create a new part to fit with existing parts in the assembly. We will see how to get an exploded view, and modify it, and how to set up section views and a drawing of the assembly. This seems like a lot, but there's actually not much involved with each topic here. Here are the sections of this lesson:

1.   Assembly Information
2.   Assembly Features
3.   Assembly and Part Modifications
4.   Part Creation in Assembly Mode
5.   Exploding the Assembly
6.   Modifying the Component Display
7.   Creating sections
8.   Assembly Drawings

To get started, make sure all the part and assembly files you created in Lesson #9 are in your working directory. Then start Pro/E and load the assembly:

File > Open > less9.asm

Shut off all the datums (planes, axes, coordinate systems), colors, and set no hidden lines. Close the model tree.

Assembly Information

In this section we will look at some Pro/E commands to dig out information about the assembly. We saw some of this way back in Lesson #1, so this will be a bit of a review. Start with

Info > Feature List > Top Level | Apply

This brings up a list very similar to the feature list of a single part. For an assembly, the list identifies all assembly features and component numbers, the ID, the name, type (feature or component), and regeneration status of everything in the assembly. Close the information window. In the **Feature List** window, select *Subassembly* and pick on the L-bracket, then *Apply*. This lists the components in the subassembly. Close the information window. Finally, select *Part* and click on the same bracket. This lists individual part features. You can see that we can dig down quite deep into the model structure. *Close* the information window and the **Feature List** window.

To see how the assembly was put together (the regeneration sequence):

Tools > Model Player

Select the rewind button to go to the beginning of the model. Proceed through the regeneration sequence with the step forward button. As you step forward, find out what information is available using the *Show Dims* and *Feat Info* buttons. Continue until you have reached the last component. Then select *Finish*.

If you want to find out more information about how the assembly was put together, in particular the placement constraints:

Info > Component

Pick on the axle. The **Component Constraints** window will open as shown at the right. This gives you a list of the placement constraints used to position the axle in the assembly. Place the cursor over one of the lines in the table - a pop-up will describe the constraint. If you click on the line, the reference surfaces will highlight in purple and blue on the model (this will be easier to see if

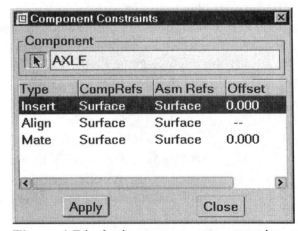

Figure 1 Displaying component constraints

you turn colors and shading off). *Repaint* the screen, select the arrow button under **Component**, and pick on another component, like one of the bolts to see similar information. *Close* the window.

Another way of looking at the logical structure of the assembly is with the model tree. Open that now in the Navigator window. Click on the small + sign in front of the *support.asm* entries. Note how the individual components are organized in levels. We used two subassemblies - their component parts are on a lower level of the tree. Add the following columns to the model tree: *Feat #*, *Feat Type*, *Status*. Reformat the column widths as in Figure 2.

Select

Settings > Tree Filters

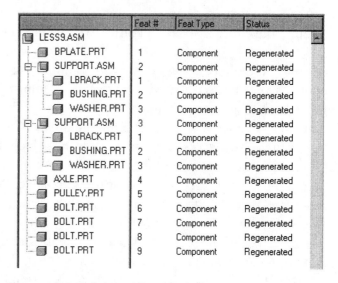

	Feat #	Feat Type	Status
LESS9.ASM			
BPLATE.PRT	1	Component	Regenerated
SUPPORT.ASM	2	Component	Regenerated
LBRACK.PRT	1	Component	Regenerated
BUSHING.PRT	2	Component	Regenerated
WASHER.PRT	3	Component	Regenerated
SUPPORT.ASM	3	Component	Regenerated
LBRACK.PRT	1	Component	Regenerated
BUSHING.PRT	2	Component	Regenerated
WASHER.PRT	3	Component	Regenerated
AXLE.PRT	4	Component	Regenerated
PULLEY.PRT	5	Component	Regenerated
BOLT.PRT	6	Component	Regenerated
BOLT.PRT	7	Component	Regenerated
BOLT.PRT	8	Component	Regenerated
BOLT.PRT	9	Component	Regenerated

Figure 2 Model tree for *less9.asm*

and turn on the display of all suppressed objects, features, and notes. *Apply* the settings and close the **Items** window. Now select one of the + signs in front of a part. The model tree shows all the features in the part. Click on any of these features and it will be highlighted on the assembly model. If you right-click on any feature, a small pop-up menu will appear with a number of the utility commands we have seen before (***Delete, Suppress, Rename, Edit***, and so on). The options are different for subassemblies. For the top level assembly (*less9.asm*), right click and select ***Info > Model***. Notice that this data has automatically been written to a file (*less9.inf*) - see the message window. The Browser controls can be used to save or print this model information. This is useful for model documentation. Close the Navigator and Browser windows.

Assembly Features

Creating Assembly Features

An assembly feature is one that will reside *only* in the assembly. You can only create them when you are in assembly mode, and they will generally *not* be available to individual parts when you are in part mode. Like features in part mode, assembly features will involve parent/child relations (either with other assembly features or with part features) and can be edited, patterned, suppressed and resumed. Our pulley-bracket assembly contains several assembly features - the default datum planes. It is usually a good idea to start a new assembly with a set of default datum planes. These would automatically be labeled **ADTM1**, **ADTM2**, and **ADTM3**, for *assembly datums*. If you use an assembly template, these default datums are created automatically and named **ASM_RIGHT**, **ASM_TOP**, and **ASM_FRONT**. The associated views are also created and saved in the view list.

Assembly features can be any type of datum (planes, curves, axes, points,...), but also solid features (extrudes, holes, ...). Open the pull-down ***Insert*** menu to see the possibilities. We will create a couple of assembly features in this lesson. The first is composed of a longitudinal cut through the entire assembly in order to show the interior detail.

In the right toolbar, select the *Extrude* button. This opens the extrude dashboard that we have seen many times before, with one addition - the **Intersect** slide-up panel - that we will get to in a minute or two. Select the *Activate Sketcher* button. For the sketching plane, pick the right face of the base plate (assuming you in the default orientation). For the **Top** sketching reference, pick the top face of the base plate. When you are in Sketcher, select your references as the lower surface of the plate (or **ASM_TOP**), the outer edge of the pulley, and the vertical datum **ASM_FRONT**. Sketch a single vertical line from the top of the pulley to the bottom of the base plate. This should be aligned with **ASM_FRONT**. Your sketch should look like Figure 3.

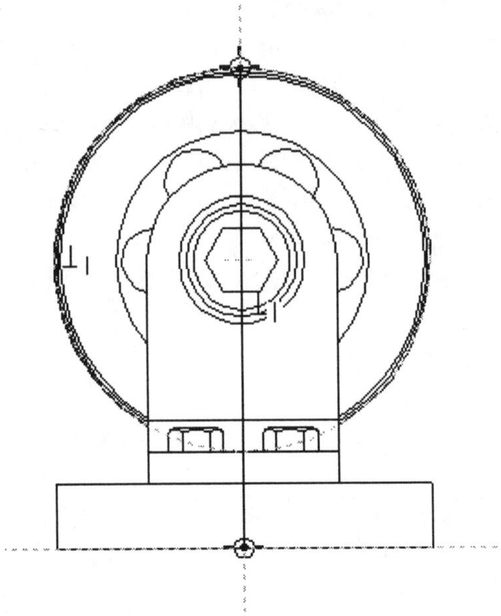

Leave Sketcher and make sure the *Remove Material* button is pressed, and the material removal side is on the left of this sketched line, and select the depth as *Thru All*. Open the **Intersect** slide-up panel. This

Figure 3 Sketch of first assembly feature (a vertical edge to create a one-sided cut)

shows which components in the assembly will be intersected by the cut. Notice that *Automatic Update* is checked (on) by default. We can remove components from the list by deselecting this option, then using the RMB pop-up menu to *Remove* selected components. Don't do that now - leave *Automatic Update* checked.

Open the other slide-up panels to see their contents - they are the same as the part mode extrude feature.

The *Preview* button on the dashboard does not work the same way as it did in part mode, so we are finished with the dashboard for now. *Accept* the cut feature. The assembly should look like Figure 4 (Why are two bolts left hanging out in space?)

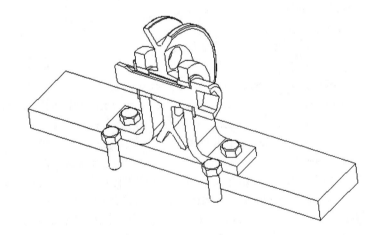

Figure 4 Cut complete

Turn the part colors back on (in the
Environment window), and shade the display.
Note that the keyway in the axle is too long -
extending into the bushings in both directions.
We will fix this a little later.

We are going to come back to the creation of
features in a little while, after a slight diversion
into display management functions.

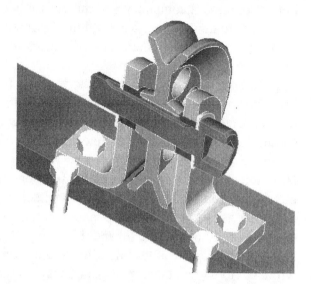

Figure 5 Cut complete - shaded view

Assembly Display Management

Pro/E can handle assemblies containing hundreds (even thousands) of components. When you
are working on such an assembly, you may want to temporarily remove some components from
the model for two reasons. First, it will remove some of the screen clutter, allowing you a better
view of just the things of current importance. If you are creating an assembly of a car, for
example, and you are working on the rear suspension, you might want to get rid of all body
panels that are blocking your view. Second, removing some (or many) components will improve
the performance of your system (faster graphics performance, faster model regeneration, less
memory required). In the car example, while working on the rear suspension, you could get rid of
the engine and steering components. Pro/E has special tools to handle cases of very large
assemblies which we won't have time to look at now. We will look at some simpler ways, that
you have already seen, to find out how they can be used to solve this problem.

Since we want to do some modifications on the keyway in the axle, we would like to remove
some of the other components from the screen that are blocking our view of this feature. We
have two options to do this: **Suppress** and **Hide**. The differences between these are as follows:

> **Suppress** - takes the component (and all the references it provides) out of the assembly
> sequence. This removes it from the screen and can speed up regeneration. Any
> children it has will be suppressed by default or will require some special handling.
> The component can be brought back into the assembly by **Resume**.
>
> **Hide** - keeps the component in the assembly sequence but makes it invisible. It can
> continue to provide assembly references (so we don't have to worry about its
> children), but must be regenerated with the rest of model. The component is made
> visible again by **Unhide**.

Let's suppress all components in the assembly except the axle and the pulley. This is not strictly necessary to do the modification, but it will remove the visual clutter from the screen. Since line color will be important here, turn the colors off.

In the model tree, select the first support sub-assembly (the one on the left in the model). Using the RMB pop-up, select *Suppress*. The sub-assembly will highlight in red and the two bolts (which reference a surface of the L-bracket) will highlight in green. Pro/E wants to know if it should suppress the bolts too. Select *OK*. Observe the entries in the model tree.

Now select the second sub-assembly - the one on the right in the model. Use the RMB and select *Suppress* as before. This component has four children: two bolts, the axle, and the pulley. Since we want to keep the axle and pulley to work on, we will have to handle them in a special way. In the **Suppress** window, select *Options*. This opens the **Children Handling** window that lists the four children (Figure 6). As you select components in this window, they will highlight in green on the model. Leave the two bolts as **Suppress**, but change the axle and pulley components to *Freeze* (click on the cell in the **Status** column and open the pull-down list or select the component and use the **Status** pull-down menu at the top). Freeze will lock the component in its current location - we can't modify its assembly position because some of its references will be missing. When the correct status has been set, select *OK*. You should now see the axle and pulley all by themselves floating above the base plate.

Children Handling	
Status Edit Info	
Object	**Status**
AXLE.PRT	Freeze
PULLEY.PRT	Freeze
BOLT.PRT	Suppress
BOLT.PRT	Suppress

Figure 6 Handling the children of a suppressed component

Notice the keyway extending past the edge of the pulley hub on both sides. In the next section, we will modify the dimension of the keyway and add some other assembly features.

Before we do that, let's see how *Hide* works. To get all the components back into the assembly, select

Edit > Resume > Resume All

Everything should now regenerate. Suppress the cut assembly feature.

Our second option for removing screen clutter is hiding components. Select one of the L-brackets either in the graphics window or in the model tree, and then pick *Hide* from the RMB pop-up menu.

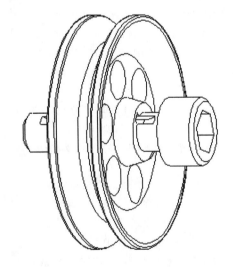

Figure 7 Assembly with hidden components

It's that easy! Notice that the Hide status applies to the L-brackets in both sub-assemblies. Also, since the component is still regenerated, we don't have to worry about losing references for its children (bushing, washer, bolts). Hiding will work well as long as the assembly is not too complex (since we are still paying for the overhead of regeneration). Hide all the components except the axle and pulley, as in Figure 7. Observe how hidden components are indicated in the model tree.

Assembly and Part Modifications

Active Components and Visibility

Pro/E gives you considerable flexibility in making changes in the assembly that involve components or features. The most common modification is to change one or more dimensions of feature(s) in a part, for example, to remove an interference. However, modifications can also include creating new features or components. Some changes will affect only the assembly, some will affect only sub-assemblies, while others will be felt all the way down to part level. Anything done at the part level (like modifying dimensions) is automatically visible in any higher level subassemblies or the top level assembly. We will see that the converse is not necessarily true. So, when you are working in the assembly, you have to be careful about what exactly you are modifying or creating:

- ▸ part features and dimensions
- ▸ subassembly features and dimensions
- ▸ assembly features and dimensions

There are two principal factors that determine the effect of what is done in the assembly. These factors involve consideration of the *Active Component* and the *Visibility Level*.

The active component is the one which is the current focus of operations. If an individual part is the active component, the next created feature will be added to the part (at the part's insert location). If a subassembly is active, the next component that is brought in (or feature that is created) will be added to the subassembly (at its insert location). By default, the top level assembly is the active component when the assembly is first created or loaded. Thus, in what we have done so far, each new component or feature was added at the bottom of the model tree as the assembly was put together. We will see in a few minutes how we can change the active component.

The concept of visibility level refers to features created in a multi-level assembly. By default, the feature (for example the long cut in our assembly) is visible only at the level where it was created. So, if you open the pulley by itself, it does not show the cut - it is visible only in the top level assembly. We will see later how to change the visibility of features like the cut so that they are visible in the part itself.

Before we do anything drastic, let's take care of the keyway dimensions in the axle. We could, of course, load the axle in a separate window and make the dimension changes there. We don't have to do that because, even in assembly mode, changes to dimensions defined in the part will be made **in the part**. Thus, these will show up if you bring up the part in **Part** or **Drawing** modes.

Let's see how that works.

Changing Part Dimensions

We need to shorten the keyway on the axle, and we
want to make this a permanent change in the part
(ie. reflected in the part file). First, we need to pick
the cut feature that created the keyway so that its
dimensions are visible. You can use Preselection
Highlighting (it may help if the **Selection Filter** is
set to *Features*) to locate this. Double-clicking on
the feature should make its dimensions visible in
yellow. Alternatively, you can find the feature in
the model tree, and use the RMB to pick the *Edit*
command. The dimensions will show up
something like Figure 8. If you need to move them
to make them clearer select an individual
dimension, and pick *Properties* in the RMB menu.
In the **Dimension Properties** window, select *Move*
or *Move Text*. Then click on the screen where you
want the dimension placed. Select *OK* and pick the

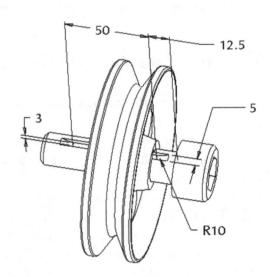

Figure 8 Original dimensions of keyway on
axle

next dimension. You can also change the *Nominal Value* of the dimension while you are doing
this.

Change the following dimensions (click on the old value and enter the new value):

> ▸   length of keyway (between the centers of the curved ends) was 50, new value = *18*
> ▸   distance from shoulder of the bolt was 12.5, new value = *20*
> ▸   radius of rounded end was R10, new value = *R5* (both ends)

Now select the *Regenerate* button in the top toolbar. Spin the axle/pulley to verify that the
keyway does not extend beyond the end of the pulley hub. To get another view of the new
keyway, select

View > Visibility > Unhide All

This will bring all the components back
into view. Now, resume the assembly cut.
If you shade the display, it should look like
Figure 9.

Figure 9 Axle keyway with new dimensions

To see what has happened to the *axle* part file, we will bring it in by itself. Click on the axle in the model tree to select it, then right click and select *Open* in the pop-up menu. The axle should show up in a new window. And, voilà, the keyway has changed. If we also had a drawing of this part and brought it up in **Drawing** mode, we would find that it has also been updated. Close the part window by selecting the X at the top right or using *Window > Close*.

Activate the assembly window with *Window > Activate* or use **CTRL-A.**

While we are dealing with part modifications, change the dimensions of the base plate. Use preselection to pick the base plate protrusion. Then use the RMB and pick *Edit*. Change the following dimensions:

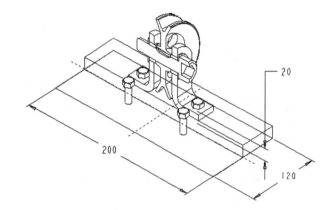

> ▸ overall length was 300, new value = ***200***
> ▸ half-length was 150, new value = ***100*** (if necessary)

The new base plate dimensions are shown in Figure 10.

Figure 10 Base Plate dimensions

Select *Regenerate*. Since these changes were made to the part, they will also be reflected in the original part file. Before we continue, suppress the assembly cut.

Adding another Assembly Feature

We are going to add a U-shaped cut to the base plate in between the L-brackets. We want to do this in the assembly so that the width of this new cut will be determined by the placement of the two L-brackets. We can't open the base plate part and create this cut since it requires references that are defined in the assembly. So, we must create the cut in assembly mode. We still have several options to deal with, each will result in a different model structure. We will look at three variations.

For our first variation, we'll create the new cut the same way we created the other assembly cut. Select the *Extrude* tool and *Activate Sketcher*. For the sketch plane, select the long front face of the base plate. For the *Top* sketching reference plane, select the upper surface of the base plate. To control the width of the cut, add sketcher references by selecting the inside vertical surfaces of the L-brackets. Make a sketch as shown in Figure 11. Accept the sketch.

Back in the extrude dashboard, for the depth, select *Thru All*. The *Remove Material* button is selected automatically. Make sure the removal direction is inside the U. Open the **Intersect** slide-up panel and deselect **Automatic Update**. Select (using CTRL) all the listed components except the base plate, then use the RMB and pick *Remove*. Notice the **Display** column for the intersected models reads *less9.asm* - our top level assembly. Also, the pull-down list for the **Default Display Level** is set to *Top Level*. Accept the feature. The assembly should now show the cut in the base plate as shown in Figure 12.

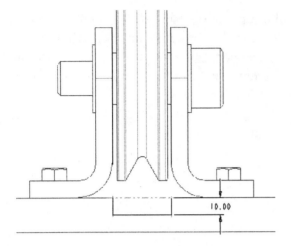

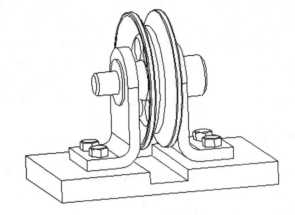

Figure 11 Sketch of second assembly feature

Figure 12 Completed cut in base plate

Go to the model tree and make sure that

Settings > Tree Filters

has checks beside all display objects. You should see that a second cut feature has been added to the assembly at the bottom of the tree. It was placed here because, by default, the top level assembly is the active component. The previous feature listed is the big cut we made before, and is currently suppressed as indicated by the small black square.

Now *Open* the base plate in its own window. You should find that the dimensions have changed (since we did that at the part level), but the new U-shaped cut does not appear in the part. Recall that the second factor which determines the extend of a modification is the visibility of the new feature. By default, our U-shaped cut was visible only in the top level assembly. Let's see how to change that. Select the *Window* command in the pull-down menus and switch back to the assembly window.

Changing Feature Visibility

In the assembly, we need to redefine the new cut so that it is visible at the part level. Select the cut in the model tree and in the RMB menu, pick *Edit Definition*. In the extrude dashboard, open the **Intersect** slide-up panel. Remove the listed component. In the pull-down list near the bottom of the panel, change the **Default Display Level** setting from *Top Level* to *Part Level*. Now pick the base plate. Notice that in the **Display** column of the table, the part is now listed instead of the assembly. *Accept* the feature in the assembly. It should look the same as before. However, open the window containing just the base plate. There is the cut! It is now visible in the part, and in any drawings of the part. Check out the feature information for the cut - this tells you where it was made.

Creating a feature in the assembly and controlling its visibility can be very useful. For example, the assembly feature might represent a custom design modification of a standard part. You may not want this modification to be reflected in the part model or its drawings (which could be used in other assemblies as well), so you would set the visibility to the top level assembly. Conversely if you want the feature to show up in the part model, set the visibility to the part. Creating the feature in the assembly also means that you can selectively affect many components in the assembly. For example, if you want to create a bolt hole through several parts, you can do that in the assembly and the holes will always line up perfectly. To make sure the hole shows up in each part, set its visibility to part level and pick all the parts you want the hole to appear in.

Select the cut in the model tree and *Delete* it. If you do this in the base plate part, it will disappear from the model tree there, but still exist in the assembly model tree. If you select the cut in the assembly, and then *Edit Definition > Intersect*, you will see that there are no intersected parts listed. This shows how you can remove a part from the visibility list of an assembly feature. Now *Delete* the cut in the assembly. There is a third option we should consider for creating the U-shaped cut.

Changing the Active Component

Suppose we want to create the U-shaped cut only in the base plate, without intersecting any other components in the assembly. Plus, we want the cut visible in the part. It makes more logical sense to add the cut to the base plate itself and not have it in the assembly at all. However, we can't create it in the part window because we need the references provided by the L-brackets in the assembly. We have to make it in the assembly, but we want it added to the part. Since the top level assembly is the current active component, any cut we create will appear in the assembly model tree. We need to change the active component to the base plate.

Select the base plate in the model tree (or in the graphics window), and use the RMB to pick *Activate*. You will be asked to confirm this (if you changed the active component by accident and without knowing, you could create quite a mess of the model!). Several changes will occur on the screen. First, in the graphics window, a label ACTIVE PART : BPLATE will appear. A more subtle change in the model tree is a small green diamond that appears on the lower right corner of the active component's icon. Third, you will note that some of the toolbar icons have changed - it doesn't make sense to have assembly-related icons in view when a part is active. This final screen change does not occur if you have set a subassembly as active.

Now, select the *Extrude* tool and create the U-shaped cut exactly as we did before. Use the L-bracket surfaces for the width references. Make a *Thru All* cut. Notice that the **Intersect** slide-up panel does not exist, since we are creating this feature in the part. Accept the feature.

Open the base plate window. Our cut is visible. Select the cut and use RMB to open the *Info > Feature* window in the Browser. This will tell you that the feature was created in the assembly. What is the implication of this? To see that, you will need to save the current assembly (this will automatically save the base plate) and erase the assembly (and all components) from the session. Notice that when you returned to the assembly window from the base plate window, the active component had reverted to the top level assembly. Now, with nothing else in session, open the base plate by itself and call up the *Info* window for the cut. Try to use *Edit Definition* and *Activate Sketcher*. Because the rest of the assembly is not in session, there are a number of missing references ("external references"). These can be a major cause of problems in complicated assemblies if file management and model organization is not well thought out. If you get into any trouble here, back out of the feature without making any changes (select the *X*).

Bring the complete assembly back into the session by opening it, and the problems with editing the U-shaped cut disappear.

What would happen to this cut if we suppressed either L-bracket? What if we used *Hide*?

To see more implications of what we have done, open the base plate in a window by itself and change the location of the holes. Referring back to Figure 6 in Lesson 9, change the dimension 52.5 to *65*. Assuming that you created the other holes by mirroring, that's all you should need to do. Now *Regenerate* the part. The cut does not change (yet!). Go to the assembly window and select *Regenerate* there. The cut widens as the two brackets follow the holes in the base plate[1]. Still in the assembly, change the dimension back to *52.5* and *Regenerate*. Can you explain the sequence of events that occurs resulting in the new display?

Part Creation in Assembly Mode

When you are working with an assembly, you may want to create a part that must exactly match up with other parts in the assembly. You could, of course, do this by creating parts separately (as we have done up to now) and by very carefully keeping track of all your individual part dimensions and making sure they all agree. You might even use relations to drive part dimensions by referencing dimensions in other part files. Here, we will find out how to create a new part using the assembly geometry as a guide and a constraint.

We are going to create the key for the axle/pulley. To simplify the environment, suppress (or hide) all the other components and assembly features except the axle and pulley (if you use *Suppress*, remember to *Freeze* these children). Turn on the datum planes and hidden lines. To create the new part, select the *Create Component* button 🔲 in the right toolbar. In the window that opens, select

> *Part | Solid | [key] | OK*
> *Create Features | OK*

Now the graphics window indicates that the active part is KEY. In the model tree, a new part *key.prt* has been added to the assembly. It is the active component (notice the green diamond). (NOTE: in the top toolbar and *File* pull-down menu, the *Save* and *Save A Copy* buttons are grayed out when the top-level assembly is not active.) The right toolbar contains the regular part-level feature creation tools. Select the *Extrude* tool, and *Activate Sketcher*. For the sketch plane, select **ASM_RIGHT**. For the *Top* reference plane, select **ASM_TOP**. It is probably better to

---

[1] The need to *Regenerate* a model twice in order for modifications to propagate through the model does occur from time to time.

use assembly datums for these references, since they are less likely to change or be suppressed. In Sketcher, zoom in on the central hub of the pulley.

In Sketcher, we will use the existing edges of the keyway in the axle and pulley to create the sketch for the key. Close the **References** window. If you get a message about trying to sketch before specifying references, you can ignore it. Select the

Use Edge

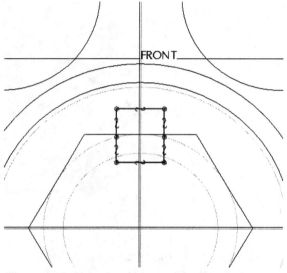

Figure 13 Sketch of key using *Use Edge*

button. Selecting existing edges will create references automatically. In the **Type** window, use the *Single* setting and click on the edges of the keyway in the axle and pulley (these are hidden); then *Close* the **Type** window. Be sure to select them all (six picks), to create a rectangular, closed section as shown in Figure 13. Note the symbol on the sketched lines that indicates they are using an existing edge.

Note that we didn't have to provide any dimensioning information for sketcher - it automatically reads the dimensions from the previous parts. This means that if we change the keyway dimensions in the pulley, the key will automatically change shape. Note that we have not explicitly connected the width of the keyway in the pulley to the width in the axle. You might think about how you could do this. What would happen if you increased the keyway width in the axle but not the pulley? How could you avoid this potential problem?

Accept the sketch. For the depth of the key protrusion, select *Symmetric*, and enter a value of *18*. Accept the feature.

Check out the model tree - you should see that *key.prt* has been added to the assembly. Make the top-level assembly active (select it and use the RMB menu). Resume the longitudinal cut to see inside the assembly (select the cut in the model tree, use the RMB and pick *Resume*). You can also select

View > Visibility > Unhide All

and shade the display. You can now see our rectangular key. See Figure 14. Hmmm... why hasn't the key been cut along with all the other parts?

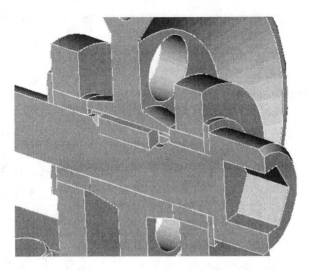

Figure 14 New *key.prt* in assembly

Now is a good time to save everything. You will find that the new part file **key.prt** is automatically saved for you - when you select *Save* in assembly mode, every object that has changed since the last save is also saved. Open the key into its own window; the only dimension shown on the part (with *Edit*) is the length. All the other dimensions are determined by the edges used in the assembly, and therefore can't be modified within part mode. What do you think would happen the next time you start Pro/E if you try to open the key part by itself?

Helpful Hint

If you are going to create parts in assembly mode, try to arrange as many *size and shape* dimensions as possible to be contained within the part. Use other assembly features only for *locational* references (like alignments, or dimensions to locate the new part). The fewer external references you have, (generally) the better.

If you load a part containing assembly references, these can sometimes be hard to track down. Let's see what we can dig out for the key. Select

Info > Parent/Child

and pick on a surface of the key (or the protrusion feature of the KEY part in the model tree). The **Reference Information Window** opens. In the left pane are the parents of the feature. These are the individual features in other parts (part names given) in the assembly (name also given). Close the **References Information** window.

With the first feature in the key part created, you can continue to add new features to the key either in its own part window, or in the assembly window. Remember that by making the key the active component, the next feature(s) you create would automatically be added to the key.

Exploding the Assembly

A useful way of illustrating assemblies is with exploded views. Creating these is very easy.
Return to the assembly window. First, suppress the longitudinal cut. Getting an exploded view is
a snap. In the pull-down menu at the top select

View > Explode > Explode View

All the parts will be translated by some default distance.
You should see something like Figure 15.

The assembly has been exploded in directions, and by
distances, determined by Pro/E. For a better view, we
can change the explosion distances. Select

View > Explode > Edit Position

We modify the exploded position of each component by
specifying a motion type and direction reference, and
then dragging one or more components in the chosen
direction. The direction can be defined by an axis, edge,

EXPLD STATE: DEFAULT

Figure 15 Default exploded view

normal to surface, and others. Make sure all the axes are displayed since we will use them to
define the explode directions. You might like to turn off the axis datum tags (with *View >
Display Settings > Datum Display*) Then, select (in the **Motion Reference** list):

Entity/Edge

and pick on the axis of the axle. The command window will instruct you to *"Select component(s)
to move"*. Click on the axle and drag it away from the L-bracket. Click again to drop it at the
new position. The component is constrained to move in the direction of the axis. Do the same
for the bushings and washers. When you are satisfied with these positions, select (again in the
Motion Reference list)

Plane/Normal

Now pick on the top surface of the base plate. Move the bolts, pulley, and key upwards away
from the base plate.

Use a combination of *Entity/Edge* and *Plane/Normal* to produce the exploded view shown
below. Of course, throughout all this, the dynamic view controls are active so you can spin and
zoom your view to your heart's content! What you cannot do is call up any of the saved views,
since these contain the explode state when the view was defined (ie, probably unexploded).
Experiment with the other options for specifying the movement direction and distance. When
you are finished, select *OK* or middle click twice.

Figure 16 Modified explode distances

Before we continue to the next section, unexplode the assembly:

View > Explode > Unexplode View

You might also like to save the assembly. All your modified explosion distances will be kept in the assembly file and will be used the next time you explode the assembly. **There is no need to create another assembly file (for example using *Save A Copy*) containing the exploded assembly.**

Component Display Style

Display styles provide another tool for dealing with the display of very complex assemblies with many components. So far, we have seen the *Suppress* and *Hide* commands to deal with the problem of screen clutter. In addition, the display toolbar buttons at the top control the display of related entities on the screen in the same way: datums are either on or off, components are either shaded, wireframe, hidden line, and so on. Here is an easy and very flexible way of setting the display of individual components in a complicated assembly. Before we start, expand the *support.asm* components in the model tree. We are going to set the display appearance of each component individually. Turn on *Shading* and colors.

Select the two L-brackets (using CTRL). Then in the pull-down menu, select

View > Display Style > Hidden Line

The two components now appear in hidden line instead of shaded. This allows us to see "through" the brackets without taking them right out of the model. The graphics window also contains the notation for the STYLE STATE : Master Style(+). We'll find out what this means in a moment.

Now select the base plate and pulley and pick

View > Display Style > No Hidden

Finally, pick the axle and the four bolts (with CTRL) and select

View > Display Style > Shaded

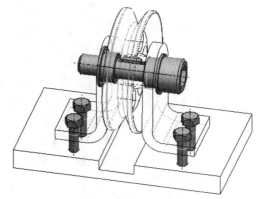

STYLE STATE: Master Style(+)

The screen should look like Figure 17. Now, use the toolbar buttons at the top (*Shaded, Hidden Line*, etc) to change the display state. These buttons now affect only the parts we have not included in the style definition (bushings, washers, and key).

Figure 17 Setting up a Display Style

We would like to set up the model so that we can get to this display style whenever we want. This is accomplished with a new tool in the top toolbar called the *View Manager* .

Select *View Manager* now. This opens the **View Manager** window (Figure 18). Pick the *Style* tab. You will see the style **Master Style(+)**. The "+" sign indicates that the master style has been modified. Highlight this and in the RMB pop-up, select *New*. This creates a new style with the definition currently in the master style. Typeover the default name (Style0001) of the new style with something like "mystyle". Notice the change in the graphics window. If you want to change back to the original Master Style display state, just double click it in the **View Manager** window. You can easily move back and forth between these two (or more!) defined display styles.

Figure 18 The View Manager

If you want to change any settings for a display style, try this: highlight our newly created *Mystyle* in the **View Manager** pane, and in the RMB menu select *Redefine* (also available in the local *Edit* pull-down menu). This brings up the EDIT window (Figure 19) which allows you to edit the display. If you select the *Show* tab, you can select a desired display state, then pick on components in the model tree or the graphics window. This also temporarily changes the contents of the model tree to show the display state of all components as defined in the selected style. In addition to the four states we have seen so far (wireframe, no hidden, hidden, and shaded) we can also *Blank* a component. This is essentially the same as hiding it. With the **Blank** tab selected, pick the key in the model tree, then *OK*. It will not disappear until you press the *Update Model* button 🔍.

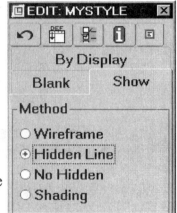

The buttons at the top of the EDIT window are *Undo*, *Reset* (to remove a style setting from a component), *Show Selected* (Navigator will show only those components that have a defined style), *Show Info* (about selected components), and *Rules* (for automatically determining a display state.

Accept the current style definitions using *Accept* in the EDIT window. This returns you to the **View Manager**. At the bottom of this window, select *Properties*. This brings up a display (Figure 20) of all the components in the assembly that have a style defined. There are buttons across the top that correspond to each possible style. By selecting a style and picking on the component you can change its style. Which button corresponds to *Blanking* the component?

Figure 19 Editing display styles

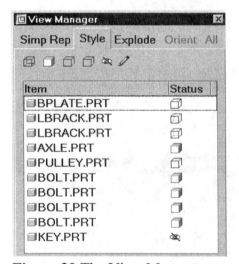

Figure 20 The View Manager Properties pane

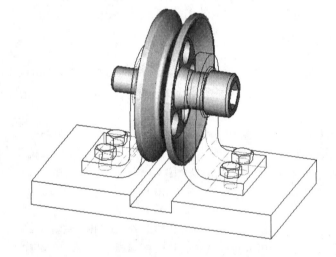

Figure 21 Component display state #2

Try out the various *Properties* options to set up another display style as shown in Figure 21.

To return to the normal display state, select the *List* button in the **View Manager**, and double click on *Master Style*.

Notice that the **View Manager** window also contains controls for the *Explode* display. Select that tab and double click on the listed style "Default (+)". In the RMB menu, select *New*, and enter a new name for the explode state "myexplode". You can easily jump between the default state and any other custom explode state just by double-clicking on the desired state. You will find that you can set some components to explode while others do not (*explode status*), explode position, etc.

Close the View Manager window and save the assembly. The display style definitions are saved with the model, just like the explode state.

Sections

Cross-sections, or just sections, are very common in assembly drawings (which we will get to next). Although sections can be created in drawing mode, it is often handy to have them available when you are working in assembly mode. This can allow you to check for fit and/or interference between components, measure clearances, and just generally help understand how an assembly fits together. Sections are very easy to create. You can define multiple sections in the same model, of different types and at different locations, and display as many of them at the same time as you like. Note that you do not have to be in an assembly to create a section - you can do it on an individual part too (and then measure section properties like cross-sectional area).

In the pull-down menus, select

> *Tools > Model Sectioning > New*

A default section name is presented. Typeover this with a new section name - "A". This now opens some of the old-style Pro/E cascading menus on the right. We will accept all the defaults here:

> *Model | Planar | Single | Done*
> *Plane*

Pick on the datum plane **ASM_FRONT**. The screen will flash and we are back in the **Sectioning** window. Double-click on the section name *A*. This opens the **Visibility** window. Select the radio button *No Clip* and turn on the toggle *Show Cross Section Geometry*, and *Accept*. The section shows up on the assembly in yellow. Note the different hatching in the different components. It is possible to change the properties of the hatching so that each component is hatched in a different color, different hatch style, and so on. Select the section "A" and in the RMB menu, select *Redefine > Hatching* to access these tools.

You might like to experiment with some of the other sectioning options. The section view will stay displayed on the model until you come back to the **Model Sectioning** window, double click on the section name (*A*), and turn off the toggle beside *Show*. Or, highlight the section, and in RMB select *Unset Visible*.

An important thing to remember here is that assembly sections can only be created on assembly datums (or make datums created in the assembly).

Save the assembly. The section definition is saved with the assembly.

Assembly Drawings

Our last task is to create a drawing of the entire assembly. We will not do any dimensioning here, just lay out the views and provide some leader notes. Select

> *File > New > Drawing > [less9asm]*

Deselect the **Use default template** option, and use an empty A-sized drawing sheet. Close the Navigator. Now select

> *Insert > Drawing View* (or select the *Add View* button)
> *General | Full View | No Xsec | Unexploded | No Scale | Done*

and pick a view center point on the left side of the sheet. For the **ORIENTATION** select

> *Saved Views | RIGHT | Set > OK*

Modify the scale of the drawing to **0.5** by double-clicking on the scale legend at the bottom of the drawing and entering the new value.

Now we'll add a section view that uses the section "A" that we defined previously. Select

> *Insert > Drawing View* (or select the *Add View* button)
> *Projection | Full View | Section | Unexploded | No Scale | Done*
> *Full | Total Xsec | Done*

Make the center point of the view to the right of the main view. Now we have to tell Pro/E what we want to section. Select *A* in the **XSEC NAMES** window. Note that we could also create new sections here as well. Now Pro/E wants to know in which view to show the section line and arrows - click on the primary view (the first view we put on the drawing).

Let's add one more view - the exploded assembly. You may have to move the two existing views down a bit to fit this one in. Then select

> *Insert > Drawing View* (or select the *Add View* button)
> *General | Full View | No Xsec | Exploded | Scale | Done*

Place the view near the top of the sheet and select our customized explode state "myexplode". When prompted for the scale, enter *0.25*. Leave the view in default orientation. You can move it around (using *Move View*) until it fits nicely.

To change the hidden line display in all the views, use CTRL and pick on the three views. Then use the RMB menu and select

>   ***Properties > View Disp***
>   ***No Hidden | No Disp Tan | Drawing Color | Done***

Select ***Done*** again to deselect the views.

We're almost finished. You should probably modify the hatching in the section view (pick the hatch, then in the RMB menu select ***Properties*** and play with the spacing, angle, and hatch pattern). You can set the hatch pattern for each component separately. Any hatch properties set in the assembly will be carried over to the drawing. This includes standard patterns for common materials (iron, steel, aluminum, and so on), hatch spacing and angle, color. When you are finished setting the hatch for a component, select ***Next Xsec*** or ***Pick Xsec*** to move on to the next one. As you would expect, the hatch styles of the components are independent of each other. Finally, add some leader notes. Your final drawing might look something like Figure 22.

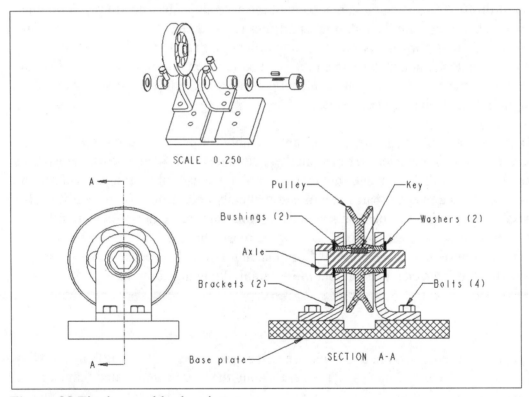

Figure 22 Final assembly drawing

This is a keeper! Obtain hard copy.

This concludes our discussion of assemblies. We have seen a number of the tools available for dealing with assemblies, assembly features, and some utilities to help you work with assemblies. What we have not had time for here is a discussion of some very important ideas concerning how to use assemblies most effectively in a design environment. You have probably heard of terms like "top-down design, bottom up implementation". Because of associativity, Pro/E is very powerful when used in this way. As you become more experienced with Pro/E, you will discover ideas about "skeleton models", "layouts", and others. The *Advanced Tutorial* presents more information on some of these topics.

Before we leave, here are some final points you should know about:

Helpful Hints

❶Using *Save A Copy* in Assembly mode is somewhat tricky. You must remember that the assembly file does not contain the parts themselves, only references to other files. In previous releases, Pro/E would automatically make duplicates of all the individual component files and append their names with an underscore "_" character. As of Pro/E 2001, when making a copy of the assembly file, you are given options about what to do with the individual part files (copy them, rename them, etc.). **Do not use *Save A Copy* in Assembly mode unless you really know what you are doing!** That is, practice this some time with parts/assemblies that you can afford to lose or damage, and when you have enough time to experiment. Don't try this the night before the big project is due.

❷ If you want to change the name of an assembly (or part) file, use the *File > Rename* function in Pro/E. For assemblies, you must do it while the assembly is in session so that the new name can be registered in the assembly. If a part is used in an assembly, renaming the part file will automatically update the assembly if it is also in session (even if it isn't being displayed). **If you copy or rename files outside Pro/E, you can expect trouble!** One of the most common error messages you will get when retrieving assemblies is "Component not found" because its name or location has been changed without the assembly knowing about it.

In the next, and final, lesson in this book, we will return to the creation of features, with an introduction to sweeps and blends. These are among the more complicated features available in Pro/E, even in their simplest forms. However, they can produce a very wide variety of geometric shapes, and should be in your modeling "toolkit."

Questions for Review

1. How can you find out the order that components are brought into an assembly?
2. How can you determine the assembly constraints used for a particular component?
3. What happens if you left click on a component in the model tree? Right click?
4. In the assembly model tree, is it possible to find out what individual features were used to create an individual component?
5. What is an *assembly* feature?
6. Can assembly features refer to individual part features, or only to other assembly features?
7. What kind of features can be created as assembly features?
8. What are the differences between using the default and the empty assembly template?
9. Can you modify individual dimensions of a part while in assembly mode? Is this a permanent modification (that is, is the part geometry changed if you load it alone)?
10. In assembly mode, how can you add a feature to a part so that it becomes a permanent feature in the part? What advantages would this have? What is the alternative?
11. If you change a feature dimension on an assembly drawing, what happens to the part containing that feature a) by itself, and b) in an assembly?
12. How can you get a cut-away view of an assembly?
13. How do you explode an assembly?
14. What does *Automatic Update* do when creating a cut through an assembly?
15. How can you explode some components and not others?
16. Draw a graphical representation of the model tree for the pulley assembly, and trace all the parent/child relations in the assembly.
17. What is contained in an assembly template?
18. If you have appropriate hardware, you may be able to set up the display so that some of the assembly components are transparent. How does this compare to the use of display styles?
19. If you want a section view in an assembly drawing, does the assembly model require a cut feature along the sectioning plane?
20. What does the command *Activate* do?
21. How do you set up a display style? What options are available? Where is the style stored?
22. Can you combine display styles and explode states in random combinations?
23. Can a model have more than one section at a time? Can you display more than one at a time?
24. How is the active component indicated in the model tree?
25. Assembly sections can only be defined using _____.
26. Under what circumstances would you set the visibility of an assembly feature to Top Level? What about setting it to Part level?
27. What is the difference between *Suppress* and *Hide*?
28. What happens if a sub-assembly is the active component when you select the *Create Component* tool?
29. Can you drag and drop the insertion point in the assembly model tree?
30. Can you reorder components in the assembly tree using drag and drop?
31. How can you change the assembly references of a component?
32. Can you select a Hidden component for a feature operation (like *Edit* or *Suppress*)?

Project

Complete the vise assembly using the parts you made at the end of each lesson. There is one more part to make - the ring shown below. It fits underneath as shown in the second figure. Create this part in assembly mode, using the existing features for dimensional references. A cutaway and exploded view are shown on the next page. The completed vise is on the cover. A VRML model is available on the web at **http://www.sdcpublications.com/tutorial**.

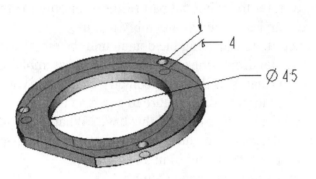

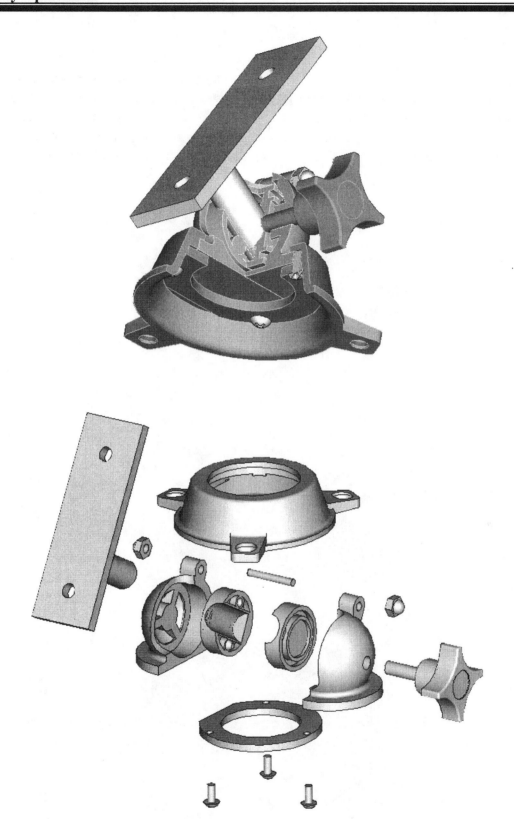

This page left blank.

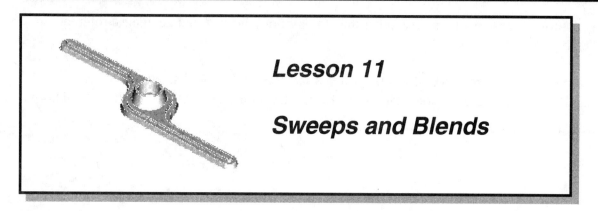

Lesson 11

Sweeps and Blends

Synopsis

Simple sweeps and blends; sketched holes; and the *Shell* command.

Overview of this Lesson

So far in these lessons, we have only used two sketched features (extrudes and revolves) to create most of our geometry. This lesson will introduce you to the simpler versions of two more sketched feature types. These can both be used to either add or remove material (protrusions and cuts) and can pack a lot of geometry into a single feature. Even these simple versions are quite complicated, which is why they have been left until last! The two features are:

Sweeps
> a feature that sweeps an open or closed section along a specified trajectory (like an extrude that follows a curved path)

Blends
> a feature that allows smooth transitions between specified cross sections (like an extrude or revolve with a varying cross section)

Sweeps and blends are advanced modeling features with many options. We will have a look at simple versions of these to create several different parts. These are totally independent of each other, so you can jump ahead to any one of these:

1. Sweeps
    - Sweeping a Closed Section
    - Sweeping an Open Section
2. Blends
    - A Straight Parallel Blend
    - A Smooth Rotational Blend

In the Wildfire release of Pro/E, only the first of these four feature variations uses the new dashboard-based interface. The other three use the old, cascading-menu style. Evidently, these will be updated to the dashboard interface in the next release. For this reason, we will not go into much detail with these features here, other than to point out the required geometric references and elements.

Along the way we will also create what is called a *sketched hole*, and discover the **Shell** command. As usual, there are some Questions for Review and Exercises at the end of the lesson.

Sweeps

There are a number of different sweep geometries available in Pro/E. We will look at just two of them: sweeping a *closed* section along an *open* trajectory, and sweeping an *open* section along a *closed* trajectory[1]. Even for simple sweeps, other combinations exist and can produce a wide range of geometries, as illustrated in the figures below. A sweep can be used to create a protrusion or cut. In the following, we will just create protrusions.

closed section, open trajectory

closed section, closed trajectory

open section, closed trajectory
(inner faces added)

Closed Section, Open Trajectory - The S-Bracket

The first part we are going to create is shown in Figure 1. The part consists of two features: the solid protrusion block at the left, and the S-shaped sweep coming off to the right.

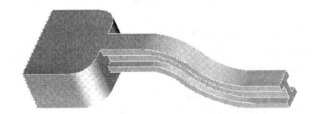

Figure 1 The S-bracket

Start a new part called **s_brack** using the default template. First create the block as a **solid protrusion, one-sided**, with a **blind** depth of **60** using **TOP** as the sketching plane and **RIGHT**

---

[1] Advanced sweeps (including *helical sweeps* and *variable section sweeps*) are covered in the **Advanced Tutorial** from SDC.

as the **Right** reference. The right edge of the sketch **aligns** with **RIGHT** and the sketch is symmetric about **FRONT**. The sketch for this protrusion is shown in Figure 2.

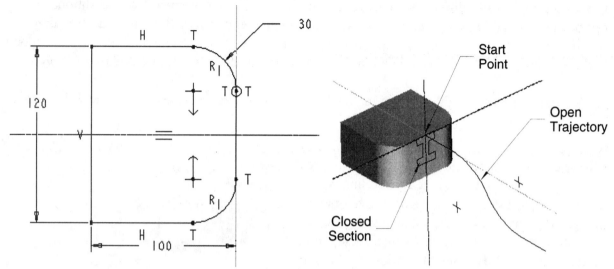

Figure 2 Sketch of the S-bracket base feature

Figure 3 Elements of a simple sweep

Now we will create the sweep. The elements of the sweep feature are shown in Figure 3. These are the *trajectory* and the *section*. The trajectory is the path followed by the section as it is swept. In this case, the trajectory is *open* and the section is *closed*. For simple sweeps, the section stays perpendicular to the trajectory. The trajectory can be either an existing edge or datum curve, or it can be sketched as we will do here. The cross section of this sweep is like an I-beam. It is created on a sketching plane located at the *start point* of the trajectory.

Defining the sweep is done in two steps: creating the *sweep trajectory*, then creating the *cross section*. The geometry of these is shown in Figures 4 and 6. We have a couple of options for the procedure we follow to define the trajectory. For either option, the trajectory is a sketched datum curve. The difference is whether we create the datum curve first and then launch the sweep command, or launch the sweep first and then select or create the curve. We will try three methods to see what effect this has on the model.

For our first sweep, we'll create the curve first.

Select the ***Sketched Datum Curve*** tool. Pick the **FRONT** datum as the sketching plane. The default **Right** reference is the **RIGHT** datum. When you get into sketcher, select the top of the block as a reference (and you can remove the TOP datum plane reference). Create the sketch shown in Figure 4.

Now we will create the sweep. The sweep creation tools are located in two

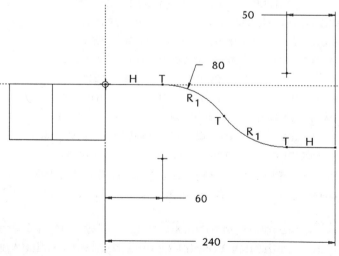

Figure 4 Sketch of datum curve (trajectory)

places in the Wildfire interface. One tool is in the right toolbar. This tool uses the new dashboard interface and gives you access to options for making advanced sweeps. We will be using this tool in our first sweep keeping all the defaults to keep things simple. The current implementation of this sweep tool in Wildfire does not cover all possible sweep geometries. Many additional options are available using commands in the pull-down menus that follow the old cascading-menu style of operation. We will see this in the next part we make. Evidently, all the sweep tools and commands are to be consolidated in the dashboard interface in a future release of Pro/E.

With the datum curve highlighted in red (as the last feature created), select the ***Variable Section Sweep*** button ![icon] in the right toolbar.

The sweep dashboard opens and the datum curve becomes highlighted with a heavy red line (with a label *Origin*). As you move your mouse cursor along this trajectory, the cursor changes style to indicate what is pre-selected. As this is happening, hold down the RMB to see the options in the pop-up menus. At each end of the trajectory are two squares and parameters "T" which we will discuss later. At one end of this trajectory, there is a yellow arrow pointing along the trajectory. This defines the *start point* of the trajectory. We want to have

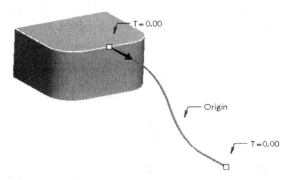

Figure 5 The sweep *Origin* trajectory

the start point at the left end as shown in Figure 5. If it is at the other end, just left click on it - it will jump to the other end. Or, you can hold down the RMB and select ***Flip Chain Direction***.

Notice the dashboard at the bottom of the screen. The default is to create a swept surface (the second button). Select the button on the far left to create a swept solid. This changes the dashboard options a bit (it adds the ***Remove Material*** option, for example). Open the ***References*** slide-up panel. The origin trajectory (our datum curve) is listed in the top pane. We will leave all the other options in this panel alone. In the ***Options*** slide-up panel, we can select either a ***Variable Section*** or a ***Constant Section***. As their name applies, this setting determines whether the feature will allow the swept section to change size/shape as it moves along the sweep trajectory. A constant section is a special case of a variable section. It requires less computation, so use it when appropriate, like now!

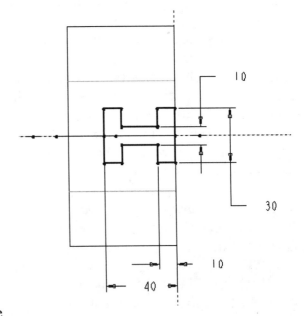

Figure 6 Sketch of sweep cross section

Now, we move on to the second step - creating the cross section. Select the ***Activate Sketcher*** button in the dashboard. The view reorients so that you are facing onto the right face of the block.

The screen should show you a yellow cross hair that automatically defines your sketch references. This is centered on the *start point* of the trajectory with the sweep coming toward you. You might like to rotate the view a bit to see the orientation of the sketch that is determined automatically by Pro/E. Use the Sketcher tools to create the cross section shown in Figure 5. The constraint display has been turned off in the figure - can you figure out what constraints are active?

When the section sketch is complete, you can select *Accept* in the Sketcher menu. The sweep will preview. If everything is satisfactory, you can *Accept* the sweep. The part should now look like Figure 7.

Note the icon for the sweep in the model tree.

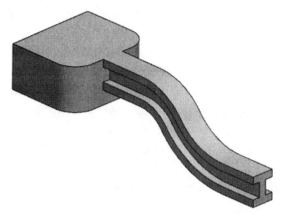

If you double-click on the sweep, you will see all the dimensions for the swept section. To change the dimensions of the trajectory, select the datum curve in the model tree and select *Edit* in the RMB

Figure 8 Completed sweep

menu. Not all combinations of dimensions are guaranteed to work, however. For example, if you increase the height of the section from 40 to 60, then 80, then 100, the feature will eventually not regenerate. Try to figure out why. (A hint is given at the end of this section!)

Finally, notice the model structure in the model tree. The datum curve and swept protrusion are separate features, with the sweep being a child of the curve. If you aren't going to use the curve for anything else, you might want to Hide it in the model tree so that it doesn't appear on the screen (as in Figure 7). This would also prevent you from accidentally picking it as a reference for another feature. It was mentioned above that changing the order of operations can affect the model tree structure. Let's see how. In preparation for the next few steps, delete the swept protrusion, keeping the curve in the tree.

Alternate Method #1 for Creating Sweep

Previously, we launched the sweep tool with the curve pre-selected (highlighted in red). This time, make sure the curve is not pre-selected (it should just be showing in blue), then select the *Variable Section Sweep* tool. The sweep dashboard opens. Follow the prompts in the message window. The first thing to do is to select curves or chains to form the origin trajectory. Pick on the blue datum curve. Make sure the start point is at the left end (click on it if necessary), select the *Solid* button, and launch Sketcher. Create the same sketch as before and accept the feature using all the defaults.

Open the model tree. There has been no change in the model structure by selecting a pre-existing datum curve for the trajectory.

Let's try one more option. This time, delete both the sweep and the datum curve.

Alternate Method #2 for Creating Sweep

The part should consist of just the block. Select the *Variable Section Sweep* tool in the right toolbar. Note that we are launching this with no trajectory curve created yet. When you are prompted to select the trajectory, launch the *Sketched Datum Curve* tool to create the origin trajectory (Figure 4). Notice what happens in the sweep dashboard - the sweep is paused while we create the curve. When the curve is created and accepted in Sketcher, we are back in the sweep dashboard. *Resume* the dashboard with the ▸ button. Then, *Activate Sketcher* and create the section as before (Figure 6). Accept the feature.

Now have a look in the model tree. This time, the curve is placed inside a group AUTO_GROUP which also contains the sweep. The curve is automatically hidden. This is the same behavior as a make datum created on-the-fly to provide a sketching plane or reference. You might think about the ramifications of this and when you should use this method instead of the previous ones.

Extending the Trajectory

You will recall that at each end of the origin trajectory were two "T" parameters. What are those for?

To see a potential problem with the current sweep definition, *Edit* the block and change the radius of the rounded corners to *55*. *Regenerate* the part and zoom in on the junction of the sweep to the block. You will see small cracks on either side (Figure 8). The sweep is obviously not meeting up with the block on the rounded surfaces.

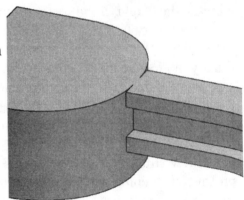

Figure 8 "Cracks" formed at junction of sweep and block

In the model tree, select the sweep and then *Edit Definition*. This opens the dashboard and shows the preview of the sweep, including the trajectory and the T-values. Pick the value at the block end of the sweep and change it to *10*. The sweep extends tangent to the existing trajectory by the new value. See Figure 9.

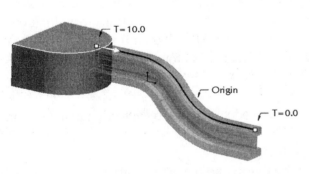

Figure 9 Extending the trajectory

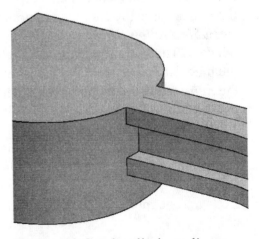

Figure 10 Cracks eliminated!

The change in the trajectory "buries" the end of the sweep in the block. If you accept the feature, the cracks will be eliminated (Figure 10).

As an alternative to extending the trajectory by a specified amount, try this. Select the sweep and **Edit Definition**. Pick the square marker at the start point and select **Extend To...** in the RMB menu. Pick on the face of the block on the opposite side of the block. This will extend the trajectory all the way across the block (and removes the T value).

To return the origin trajectory to its original shape, you need to select the curve again. Open the **References** slide-up panel. Click on the listed Origin trajectory and then one of the rounded edges of the block. This creates a trajectory for a sweep following that edge. Now pick on the original datum curve. The sweep origin trajectory is now defined the same as it was before. Extend it using the T parameter and accept the feature.

Save the part (for future experimentation!).

Before we leave this sweep, you should note the following:

- It is not strictly necessary for the cross section to lie exactly on the trajectory. If the section is offset from the trajectory at the start point, then the sweep will be offset.
- You have to be careful that during the sweep, the cross section doesn't pass through itself - this can occur when the radius of a trajectory corner is very small (relative to the section size), and the section is on the inside of the curve.
- You can sweep a closed section around an open or closed trajectory.
- The trajectory need not be formed of tangent edges. If there are corners in the trajectory, for some sweeps Pro/E will produce mitered corners in the solid, as shown at the right.
- The trajectory can also be formed as a three-dimensional spline.
- The trajectory does not have to be a datum curve - any chain of edges will do (depending on the type of sweep).

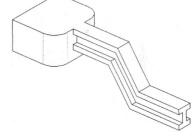

Figure 11 Mitered corners

Let's move on to the next type of sweep. Close the **s_brack** window and erase it from session.

Open Section, Closed Trajectory - The Lawn Sprinkler

This version of the sweep command will be used to create the part shown in Figure 12. This part has only three features: the sweep used to create the base with two arms, a revolved protrusion to create the hub, and a sketched counter-bored hole down the central axis of the hub. A detailed view of the arm cross section is shown in Figure 13. The stepped contour is the same all around the base of the sprinkler - this is what we will create as a sweep. The options to do this type of sweep have not yet been ported to the Wildfire interface, so we will have to use some of the old-style cascading menus.

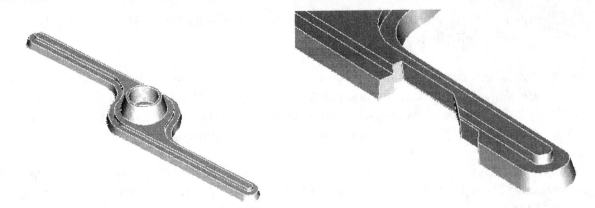

Figure 12 The Lawn Sprinkler   **Figure 13** Close-up of lawn sprinkler cross
                                                                                      section

Start a new part called **sprinkler** using the default template. The first feature we will create is the
base with the extending arms, using a sweep. As before, we do this in two steps: first the sweep
trajectory (Figure 14), then the swept section (Figure 15). The trajectory is a closed curve while
the section is open. For the section we only need to create an open curve showing the stepped
edge detail. We will use a special command to fill in the surfaces between the open swept edges
at the top and bottom of our sketch. The trajectory curve will be created as part of the sweep
itself rather than using a separate datum curve (which would also be possible).

Insert > Sweep > Protrusion[2]

The **Elements** window is now open on the right. In this window we can keep track of the sweep
elements (trajectory and section) as they are being defined. To do that, we make picks in a series
of cascading menus below the Elements window (usually accepting defaults). As you go through
here, keep an eye on the message window at the bottom for prompts. First, select

Sketch Traj

With *Plane* highlighted (notice that we could create a make datum here), select **TOP** as the
sketching plane (view direction downwards), and then the **RIGHT** datum as the **Right** reference
plane. This should bring you into Sketcher where you can create the curve shown in Figure 14.
This trajectory is for the outer edge on the bottom surface of the part. The swept section will be
inside and above this trajectory. Unless you are very good with Sketcher, don't try to sketch this
all at once. Sketcher allows you to cycle through the draw - constrain - dimension - modify -
draw sequence as often as you wish. Build the sketch up in stages. For example, start with a
central circle, add one arm, trim away the circle, add the other, trim the circle, and so on. Notice
how few dimensions are actually required to define the sketch. Try to create these as strong
dimensions and constraints as soon as possible while you are sketching.

---

[2] Notice that only *Protrusion* and *Thin Protrusion* commands are available at this time.
Normally, *Cuts* and *Thin Cuts* are also available - they aren't at this time since there is nothing
to remove material from.

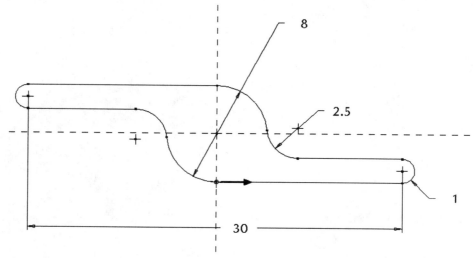

Figure 14 Lawn sprinkler sweep trajectory (Sketcher constraint display turned off)

Notice the location of the start point (with the arrow). To change the start point, select the desired vertex and then pick Start Point in the RMB menu. When the curve is created, select *Accept*.

We now move on to create the cross section of the sweep. We only have to sketch the stepped edge detail with an open curve. We will tell Pro/E to fill in the top and bottom surfaces of the part from the free ends of the sketch. Therefore, select ("*Add Inner Faces*")

Add Inn Fcs | Done

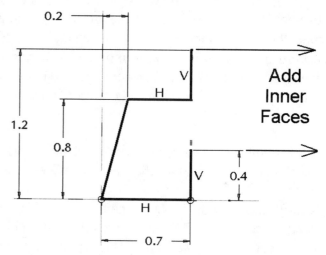

Figure 15 Lawn sprinkler sweep section

Again, you are presented with an edge view of the trajectory, with the yellow cross hairs to show where you will create the section (its sketching plane). You might have to spin the view a little to get a better idea about the orientation of the part. Sketch the open line shown in Figure 15. Compare this sketch to the cutaway view of the sprinkler back in Figure 13. You can see where the inner faces will be added, and why this only works for closed trajectories.

The purpose of the cross hair is to show you the relative position of your sketch and the trajectory. As stated above, the cross section does not necessarily have to touch the trajectory. The free ends of the cross section will be closed in by the inner faces of the sweep. When the sketch is finished, select *Accept*. You can ignore the WARNING in the message area about the open ends. Finally, in the **Elements** window you can *Preview* the feature. Select *OK* if it is satisfactory. The part should look like Figure 16.

Figure 16 Completed sweep

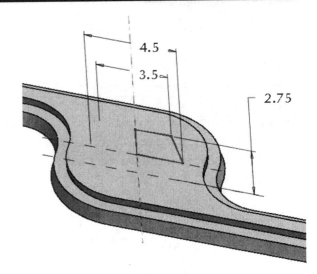

Figure 17 Dimensions of hub

Open the model tree. This sweep is entirely contained (trajectory plus section) within a single feature. If you double-click on the lower edge of the sweep (the trajectory), you can edit the trajectory dimensions. If you want to change dimensions of the section, you have to select the feature and then **_Edit Definition_**. In the **Elements** window, select **_Section > Define > Sketch_**. This takes you back into Sketcher where you can do anything with the section sketch. Hopefully, changing section dimensions will be more consistent with the Wildfire interface in the next release.

Add the hub as a revolved protrusion, sketched on FRONT and using the dimensions in Figure 17. Note that the height of the hub is measured from the bottom surface of the sweep.

Finally, create a counter-bored hole on the axis of the hub. This time, instead of a straight or standard hole, we will specify a cross sectional shape for the hole, including the counterbore. This is called a *sketched hole*. This type of hole is essentially a revolved cut that is automatically revolved through 360 degrees. We provide the cross sectional shape of the hole using Sketcher. The placement references are the same as a straight hole. Normally, users will use either a *Coaxial* placement or define a placement point.

Select the **_Hole_** tool in the right toolbar. The **Hole** dashboard opens. Change the option in the pull-down list from **_Simple_** to **_Sketched_**. To the right of this, a new button appears 🔲 - this is a variation of the **_Activate Sketcher_** button. Just to the left of this is a button that allows you to read in a previously created hole profile (say for a specially shaped bit that you use frequently). Select the **_Activate Sketcher_** button now. In the Sketcher window that opens up, create the sketch shown in Figure 18. You must create the centerline and you must also close the sketch down the centerline. When you accept the sketch, you're back in the hole dashboard. Select the axis of the hub for the primary reference. The **Coaxial** placement type is then automatic. The previewed hole will appear as a cylinder. Open the **Placement** slide-up panel to see this. Click in the **Secondary References** pane and then select the top surface of the hub. The shape of the hole is now shown. If necessary, pick the **_Flip_** button beside the primary reference so that the hole goes in the proper direction. Open the **Shape** slide-up panel - it shows the shape of the sketched hole.

The placement plane is the top surface of the hub.

Pro/E will take the top edge of the sketched hole and align it automatically with the placement plane, with the axis of the hole coinciding with the axis of the hub. We also could have used a linear placement using the datum planes but this means that if the hub moved, the hole would not go with it. Accept the hole feature.

That completes our lawn sprinkler part. Save it now.

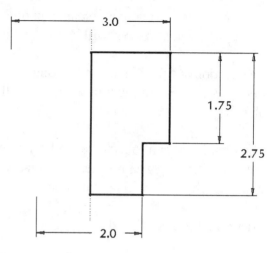

Figure 18 Sketched hole profile

So, that's the end of sweeps! As you can see, these are quite complicated features, packing a lot of geometric information into a single feature. You might like to go back and edit any of the dimensions of the sweeps to see what happens. You can modify either the trajectory, or the section, or both! Be aware that arbitrary modifications might make the sweep illegal, so save your part before you try anything drastic. If you are using the dashboard interface, it is very easy and convenient to experiment with different options since you always have a preview of the feature on the screen - if the sweep is illegal, it will not preview.

Close the sprinkler part and remove it from the session.

Blends

A blend is like an extrusion with a changing cross section. The different cross sections are specified using a number of sketches. To create the blend, the distance between cross sections is then specified. This can be either a linear distance (forming a *parallel blend*) or an angular distance (forming a *rotational blend*), or a combination of these (a *general blend*). In the following, we will look at the first two of these. A blend can be used to create a protrusion or a cut. Some restrictions apply:

- ♦ At least two sections are required.
- ♦ Each section must be created separately and constrained to either the existing geometry, a previous blend section, or a local sketched coordinate system.
- ♦ Each section must have the same number of vertices; normally this means the same number of line (or arc) segments. This rule can be overridden using a *blend* vertex (see the on-line help for information on this).
- ♦ Each section has a starting point (one vertex on the sketch) - these must be defined properly on all the sections or else the resulting geometry will be twisted.
- ♦ For a rotational blend, the section planes can be no more than 120 degrees apart.

♦   For a rotational blend, a coordinate system is needed in the sketch of each section, whose Y-axis will be the axis of rotation of the blend.

The sections of the blend can be connected either with *straight* (ie. ruled) surfaces, or with *smooth* surfaces. In the following, we will create two parts that illustrate the basic features of blends.

The blend creation tools have not been updated to the new Wildfire interface. Instead, they use the old cascading-menu style. Presumably, this update will be done in the next release. The essential elements will be the same.

Straight, Parallel Blend

This is the simplest form of a blend. We will create the part shown in Figures 19 and 20. This blend has three sections: a square, a rounded rectangle, and a final thin rectangle. These are seen best in the wireframe view of Figure 20.

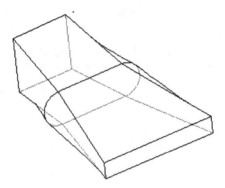

Figure 19 A straight, parallel blend   **Figure 20** Straight, parallel blend - wireframe

Start up a new part called **blend1** using the default template. To create the blend:

> *Insert > Blend > Protrusion*
> *Parallel | Regular Sec | Sketch Sec | Done*
> *Straight | Done*

Follow the prompts in the message window. We will select defaults as much as possible. Select **FRONT** as the sketching plane, and **RIGHT** as the *Right* reference plane. The sketcher references (TOP and RIGHT) are chosen for you automatically.

Each section of the blend is sketched separately (although eventually all sections appear in the same sketch). This includes dimensioning, aligning, regenerating, and so on. When each section is completed, we will move on to the next section with a special command. Do NOT select *Accept* in Sketcher until all sections have been defined. For a parallel blend, when we move on to the next section, the previous section will remain displayed on the screen in gray. The new sections can use the old ones for constraint references, or they can be defined with respect to other part features.

The first section is a 10 X 10 square centered on the references, as shown in Figure 21. Note the round dot and arrow on one of the four vertices (on the figure, it is in the upper left corner). This is called the *start point* and shows the direction that vertices will be traversed in the section. Since the square has four vertices, each section we produce must also have four, corresponding to each other in number and in sequence (clockwise here) starting from the start point. If you make an error with the start point on any of the sections, your blend will become twisted. If your sketch's start point is not in the position shown, left click on the desired vertex, then hold down the RMB and select ***Start Point*** from the pop-up menu.

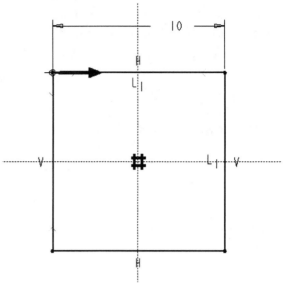

Figure 21 Blend section #1

When you have a successful regeneration, **DO NOT** select *Accept* since this indicates that *all* the blend sections have been created[3]. Instead, hold down the RMB and select

Toggle Section

This will take you to the next section (the rounded rectangle). The previous section is grayed out, and Sketcher is now used to create the second section. The sketch is shown in Figure 22. When that one is regenerated successfully, toggle to the third section (the thin rectangle). The dimensions and placement of all the sections are shown in Figure 23. Make sure all your start points are located correctly. (Use the right mouse button to get the pop-up menu.)

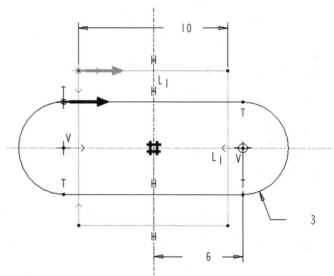

Figure 22 Blend section #2

---

[3] If you try to leave Sketcher with only one section defined, you will get an error message. After the second section, if you accidentally leave Sketcher too early, in the Elements window select **Section** in the elements list, then click the ***Define*** button and select ***Sketch***.

If you need to go back to a previous
section, use the RMB pop-up menu
to select **Toggle Section**. You can
then cycle through each of the
sections to make corrections using
Sketcher. The active sketch is
shown in yellow. When the third
section is complete, select **Accept**.

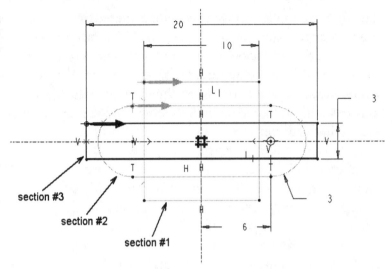

Figure 23 Completed sketch showing three sections

Now you will be asked (see the message window) for the distance between each planar section.
The distance from the first section (the square) to the middle section #2 is **15**. The distance to the
next section #3 is **20**. This should complete the specification of the blend. **Preview** the part, and
select **OK** when you are satisfied with the part.

Open the model tree and note the icon that
indicates this is a blend feature.

You might like to try to **Edit** the dimensions of the
cross sections. When you select the feature (either
in the model tree or by double-clicking on it in the
graphics window), you will see all the section
shapes with their dimensions displayed on the
original sketching plane, and the distances between
planes shown normal to the sketch as in Figure 24.

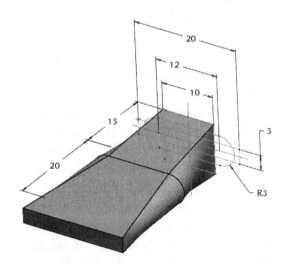

Figure 24 Completed blend showing all
dimensions with **Edit**

The *Shell* Command

Just for fun, here is a feature creation command we haven't mentioned before. Pre-select the
blend feature (edges highlighted in red) and in the right toolbar, pick the **Shell** tool 🔲 . This
defaults to hollowing out the highlighted object (shelling it) using a default thickness for the
remaining shell wall (see the dashboard). The interior edges are shown in preview yellow. A flip
button allows you to create the shell on the inside or outside of the model.

In addition to shelling the model, we want to remove the two end surfaces. Open the **References** slide-up panel. This contains two panes. Click in the **Removed Surfaces** pane and pick on the front and back surfaces (using CTRL) as shown in Figure 25. These will highlight in red as they are picked. *Preview* the feature. The designated surfaces have been removed. The other pane in the **References** panel allows us to set different thicknesses for different regions of the shell. Accept the shell feature. The part looks like Figure 26. Open the model tree and observe the icon used to indicate the shell.

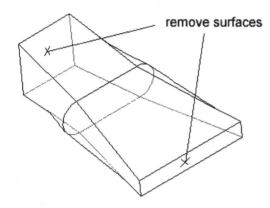

Figure 25 Surfaces to be removed

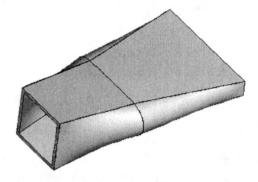

Figure 26 The *Shell*ed part

The Shell command can handle quite complicated geometry, but is not foolproof. You may find that the command will fail sometimes if the geometry is too complicated. Normally, the shell command is used fairly early in the model before this situation can develop.

Let's move on to our last new feature in this lesson. Save the current part (for later experimentation) and remove it from session.

Smooth, Rotational Blend

A rotational blend is set up by specifying the cross sectional shape on a number of sketching planes that have been rotated around a common axis. The usual restrictions apply as to the number of vertices in each section and the start point. Consecutive sections can be no more than 120 degrees apart.

Rotational blends are another feature that has not migrated to the new interface style. Maybe next time...!

We are going to make the part shown in Figures 27 and 28. Note that the surfaces on the blend are smooth, except for the two end surfaces.

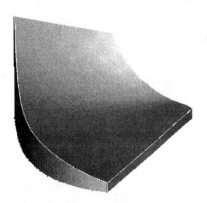

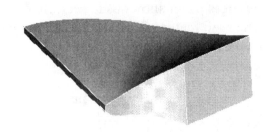

Figure 27 Smooth rotational blend - front isometric

Figure 29 Smooth rotational blend - rear isometric

If we select straight surfaces, we will get the shape shown in Figure 29. This also shows where the four sections of this blend are going to be created. Because these sections are on different planes, we will find that, unlike the parallel blend, you will define each sketch by itself in its own sketch window. Also unlike the parallel blend, you move from one section to the next using *Accept* in the Sketcher toolbar instead of *Toggle Section* as we did before. What ties the sections together is a coordinate system that we create in each section sketch.

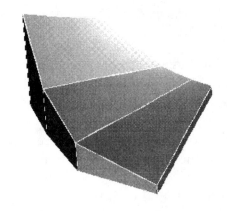

Figure 28 Straight rotational blend

Start a new part called **blend2** using the default template. You can delete the default coordinate system. Then start the blend creation:

> *Insert > Blend > Protrusion*
> *Rotational | Regular Sec | Sketch Sec | Done*
> *Smooth | Open | Done*

Select **FRONT** as the first sketching plane (the view direction is okay), and **RIGHT** as the *Right* reference plane. We are going to create four cross sections, with a separation of 30 degrees between each section. Therefore, the total angle of rotation of the blend will be 90 degrees. Each section must include a coordinate system in the sketch (discussed below). The rotation will occur around the Y-axis of this system.

The first section is a square sketched directly on **FRONT**. The lower edge is along **TOP**. The dimensions of the section (and its position on the final blend), are shown in Figure 30.

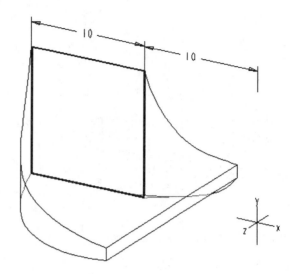

Figure 30 Rotational blend - section 1

Draw and dimension the sketch. Before you toggle to the next section, you must create a coordinate system in the sketch. This is done using the *Reference Coordinate System* tool in the right toolbar that looks like ⌙. This is on the *Sketched Point* flyout. Do not confuse this with the *Datum Coordinate System* tool. Place the sketched coordinate system to coincide with the origin of the datum planes (where the existing references cross). Notice the direction of the Y-axis.

Take note of the start point of the sketch and correct it if necessary (we want the top left corner). Since we are moving on to a different sketch plane for the next section, select *Accept*. You will be asked for the "y-axis rotation angle" to the next section. Enter *30*.

A new sketcher window opens up. In this window, you need to sketch the second section and supply a sketch coordinate system as before. Pro/E will automatically align this system with the one in the first section. The blend "rotation" is around the Y-axis of the system. Dimension the sketch to the coordinate system and make sure the start point is on the correct vertex (top left corner). The second section has the dimensions shown in Figure 31.

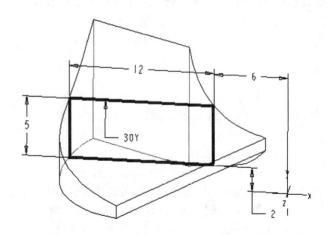

Figure 31 Rotational blend - section 2

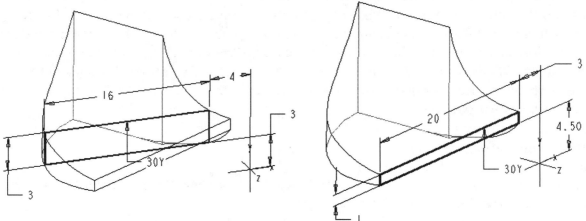

Figure 32 Rotational blend - section 3

Figure 33 Rotational blend - section 4

When you select *Accept*, proceed on to the next (third) section. The rotation angle is again 30 degrees. The third and fourth section dimensions are shown in Figures 32 and 33 above.

When the fourth section is complete and you are asked to continue to the next section, type in *n*. The message window should indicate that all elements are complete, and you can *Preview* the part.

If your start points aren't correct on any section (the blend will be twisted or may even fail), highlight **Section** in the Elements window, then click on *Define*. Follow the message window instructions. Eventually you will get to the Sketcher menu. Select the desired vertex and use the right mouse pop-up menu to set the start point.

Leave Sketcher and select *OK* in the elements window when you are satisfied. Try to *Edit* dimensions in the blend by double-clicking on it in the graphics window. The feature appears as shown at the right. Note the alignment of the coordinate systems embedded in each section sketch.

That completes our limited presentation of blends. As you can see, blends contain a lot of geometric information and are therefore a bit more difficult to set up. However, they offer considerable flexibility and can create very complex shapes not attainable with the simpler features. There are advanced features (swept blends and helical blends, for example) that offer even more complexity/flexibility. Consult the on-line help for information about these.

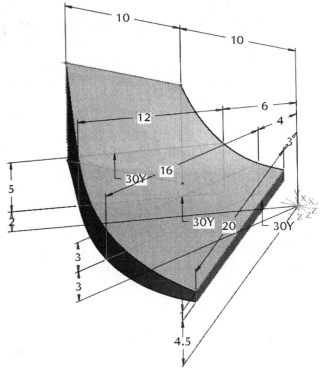

Figure 34 Completed rotational blend

Just for fun, let's try a variation of the *Shell* command on this rotational blend using a small negative shell thickness. Pre-select the two end surfaces (using CTRL), then pick the Shell tool. Enter a thickness of 0.25 and then pick the Flip button. Accept the feature. This creates a feature by adding material to the outside of the model, then removing everything from the original surface inward. This is like making a mold or negative image of the part. This will not always work (for example if the part has a very complicated geometry), and is not really the best way to make molds for parts anyway.

Save the part and close it.

This concludes our study of blends. It will be interesting to see how these are handled when they become incorporated into the new interface style.

Conclusion

Well, we have reached the end of this series of Pro/ENGINEER lessons. We have gone over the fundamentals of creating basic parts, assemblies, and drawings. Much of the material has been presented only once. It is likely that you will have to repeat some of these lessons to get a better grasp on Pro/E, and it is certain that you will need much more practice to be proficient. In some instances, we have only scratched the surface of Pro/E functionality and it is up to you to explore deeper into the commands and options. The more you know and are comfortable with, the easier it will be to perform modeling tasks with Pro/E. You may find that you will also begin to develop a different way of thinking about part design. As your modeling tasks get more complex, the need to plan ahead will become more important. You should also remember that what we have covered is only the first step in the integrated task of design and manufacturing. From here, you can head off in a number of directions: engineering analysis using Finite Element Modeling, mechanism kinematics and dynamics, manufacturing analysis, mold design, sheet metal operations, piping layout, and much more. Good luck on your journey and have fun!

Questions for Review

1. Draw a 3D sketch of an example of each of the following sweeps:
    ▸ closed section, open trajectory
    ▸ closed section, closed trajectory
    ▸ open section, open trajectory
    ▸ open section, closed trajectory
    What additional information will be required to create each (or any!) of these features?

2. Does at least one of the vertices of the swept section have to align with or be on the sweep trajectory?

3. What problem may arise if the swept section is "large" and the sweep trajectory has a "small" radius arc in it?

4. What happens if the sweep trajectory has discontinuities (kinks) in it rather than being composed of smooth "tangential" transitions?

5. What is the difference in the model between creating the trajectory datum curve first and then launching the sweep tool, or doing these in the reverse order?

6. When first entering the Sketcher window to define the section for a sweep, it is often difficult to understand the orientation of the view. How can you determine the location and orientation of the section with respect to the trajectory?

7. Find out what happens if you put the start point in the middle of an open trajectory.

8. In the exercises above, the swept section was normal to the trajectory. Is it possible to create a sweep where the section is oriented at an angle to the trajectory?

9. What does the sweep "T" value do? What is an alternative, and what does it do?

10. Can you change the geometry of the trajectory independently of the geometry of the section?

11. What is meant by "inner faces" of a sweep?

12. Can a closed section have any inner faces?

13. Can you have several non-overlapping closed sections in a single sketch for a sweep?

14. Can a sweep trajectory intersect itself (like a figure-8)?

15. What are the essential common characteristics of all sections in a parallel blend?

16. When creating a blend, what is meant by the "start point" of the sketch?

17. What is meant by a ruled surface?

18. In a straight, parallel blend must the sections all be centered on a common axis or point?

19. Are all parallel blends symmetrical?

20. In a parallel blend, do the sections have to overlap in the sketch?

21. What are the essential common characteristics of all sections in a revolved blend?

22. How can you change the dimensions of a section in a sweep or blend?

23. What happens if you try to delete one of the sections in a blend?

Exercises

Here are some parts to try out using the commands in this lesson.

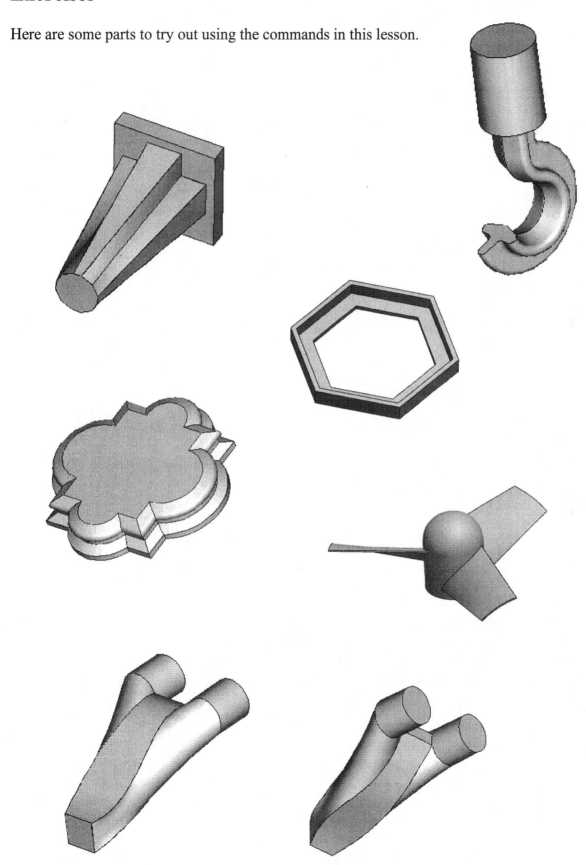

This page left blank.

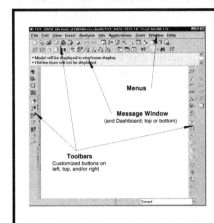

Appendix

Interface Customization Tools

Synopsis

Configuration settings; customizing the screen toolbars and menus.

Overview

This appendix describes tools for customizing your Pro/E working environment. The major customization tool is the configuration file (default: *config.pro*). We'll also look at ways for managing and creating your own custom toolbars.

Configuration Files (*config.pro*)

By now, you should be familiar with the commands for environment settings that are available in

> *Tools > Environment*

These aspects of the Pro/Engineer working environment (and much more!) can also be controlled using settings stored in configuration files (*config* files for short). Pro/E has several hundred individual configuration settings. All settings have default values that will be used if not specifically set in a *config* file.

The most important *config* file is a special file called *config.pro* that is automatically read when Pro/E starts up a new session. You can also read in (and/or change) additional configuration settings at any time during a session. For example, you may want to have one group of settings for one project you are working on, and another group for a different project that you switch to during a single session. In this tutorial, we will deal only with the use of the single configuration file, *config.pro*, loaded at start-up.

Several copies of *config.pro* might exist on your system, and they are read in the following order when Pro/E is launched:

♦   *config.sup* - this is a protected system file which is read by all users but is not available for modification by users. Your system administrator has control of this file.

♦   Pro/E loadpoint - this is read by all users and would usually contain common settings determined by the system administrator such as search paths, formats, libraries, and so on. This file cannot normally be altered by individual users.

♦   user home directory - unique for each user (Unix)

♦   startup directory - the working directory when Pro/E starts up. To find where this directory is, select *File > Open* and observe the directory name in the top box[1]

Settings made in the first copy (*config.sup*) cannot be overridden by users. This is handy for making configuration settings to be applied universally across all users at a Pro/E installation (search paths for part libraries, for instance). An individual user can modify entries in the last two copies of *config.pro* to suit their own requirements. If the same entry appears more than once, the last entry encountered in the start-up sequence is the one the system will use. After start-up, additional configuration settings can be read in at any time. These might be used to create a configuration unique to a special project, or perhaps a special type of modeling. Be aware that when a new configuration file is read in, some options may not take effect until Pro/E is restarted. This is discussed more a bit later.

Settings in *config.pro* are arranged in a table. Each row is composed of two entries in the following form:

config_option_name   config_option_value

Option values can be composed either of text, single numbers, or series of numbers. A listing and description of many *config* options is contained in the on-line help. Select the following (starting in the pull-down menus):

Help > Help Center

Then pick the links:

Fundamentals > Pro/ENGINEER Fundamentals

In the **Contents** pane, expand the topics

⊕ *Pro/ENGINEER Fundamentals*
⊕ *Configuring Pro/ENGINEER Fundamentals*

Although this shows quite a long list, it does not include every possible option (by a long shot!). Fortunately, as we will see in a minute, the dialog window for working with configuration files contains a one line description of the options. There is also a search capability for finding option names. Although this makes finding the options much easier, you are encouraged to explore the

---

[1] In Windows, right click on the Pro/E icon on the desktop (if it exists), select **Properties > ShortCut** and examine the **Start In** text entry field.

on-line help - you might find just the setting you need to make your life easier!

Your system may have a standard configuration file available for you to use as a basis for your own work. Look for the *config.pro* file in the **pro_stds** ("standards") directory in the Pro/E installation.

Before we proceed, if you have access to this file, copy it to your start-up directory, along with the file *config.win* (this is a file containing customized screen layout settings which are discussed later). Now launch Pro/E, or if it is already up erase everything currently in session and set your working directory to your normal start-up directory.

The Configuration File Editor

You can access your current configuration file using

> *Tools > Options*

This brings up the **Options** window. If your system has options set already, these will appear in the window. If not, the central area of the window will be blank, as in Figure 1. We'll discuss the operation of this dialog window from the top down.

The **Showing** pull-down list at the top will let you choose from a number of configuration groups (Current Session, your start-up config.pro, or elsewhere). Select *Current Session*.

Deselect the check box just below the **Showing** pull-down box. After a couple of seconds, a complete list of all the Pro/E

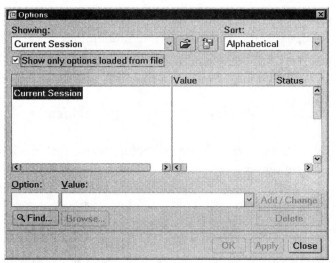

Figure 1 The Options window for setting and editing the configuration file

configuration options will appear. The first column shows its name, and the second column shows its current value. An entry with an asterisk indicates a default value.

Note that you can resize the column widths by dragging on the vertical column separator bars at the top of the display area. At the far right is a long (scrollable) one-line description of each option.

Browse down through the list. There are a lot of options here (over 750!). Note that the options are arranged alphabetically. This is because of the setting in the **Sort** pull-down menu in the top-right corner. Change this to *By Category*. This rearranges the list of options to group them by function. For example, check out the settings available in the **Environment** and **Sketcher** groups. Fortunately, there are a couple of tools to help you find the option name you're looking for. Let's see how they work.

Check the box beside "Show only options loaded from file" and select *Sort(Alphabetical)*. Note that the options listed here are only those that are different from the default settings.

Adding Settings to *config.pro*

Assuming you have a blank *config.pro*, let's create a couple of useful settings. At the bottom of the **Options** window are two text boxes for entering option names and values. If you know the name of the option, you can just type it in to the first box. For new users, a useful setting is the following. In the text box below **Option**, enter the option name *prompt_on_exit*. As you type this in, notice that Pro/E anticipates the rest of the text box based on the letters you have typed in. After typing enough characters (up to the "x" in "exit"), the rest of the desired option will appear; just hit the **Enter** key. In the pull-down list under **Value**, select *Yes*. Note that the option name is not case sensitive and the default value is indicated by an asterisk in the pull-down list. Now select the *Add/Change* button on the right. The entry now appears in the data area. A bright green star in the **Status** column indicates that the option has been defined but has not yet taken effect.

Now enter a display option. The default part display mode in the graphics window is **Shaded**. Many people prefer to work in hidden line mode - let's make it the default on start-up. Once again, we will enter the configuration option name and pick the value from a drop-down list. The option name and value we want are

> **display**   **hiddenvis**

Now select *Add/Change* as before (or just hit the Enter key after typing the "h"). Add the following option to control how tangent edges should be displayed

> **tangent_edge_display   dimmed**

Another common setting is the location of the Pro/E trail file. As you recall, the trail file contains a record of every command and mouse click during a Pro/E session. The default location for this is the start-up directory. Theoretically, trail files can be used to recover from disastrous crashes of Pro/E, but this is a tricky operation. Most people just delete them. It is handy, therefore, to collect trail files in a single directory, where they can be easily removed later. There is an option for setting the location of this directory. Suppose we don't know the configuration option's specific name. Here is where a search function will come in handy.

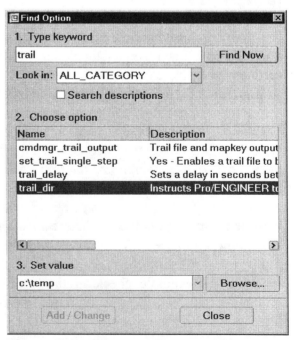

At the bottom of the **Options** window, click the **Find** button. This brings up the **Find Option** window (Figure 2). Type in the keyword *trail* and select

Figure 2 Finding a configuration file option

Look in(ALL_CATEGORY) > Find Now

Several possibilities come up. The option we want is listed as **trail_dir** - scroll the description to the right to confirm this. Select this option and then pick the *Browse* button at the bottom to identify a suitable location on your system for the value. Perhaps something like *c:\temp*. Then select *Add/Change*. The new entry appears in the **Options** window. In the **Find Option** window, select *Close*.

For some options, the value is numeric (eg setting a default tolerance, number of digits, or the color of entities on the screen). In these cases, you can enter the relevant number (or numbers separated by either spaces or commas). For example, under **Option**, enter the name **system_hidden_color**. Then under **Value**, enter the numbers **60 60 60** (separated by spaces). These give the values of red, green, and blue (out of 100). Equal values yield gray; this setting will brighten the hidden lines a bit from the default value. Select *Add/Change*.

We have now specified four options. To have them take effect, select the *Apply* button at the bottom. The green stars change to small green circles in the Status column.

For practice, enter the options shown in Figure 3. The order that the configuration options are declared does not matter. Feel free to add new settings to your file (for search paths, libraries, default editors, default decimal places, import/export settings, and so on).

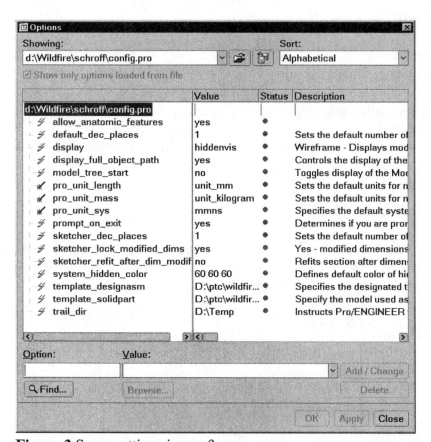

Figure 3 Some settings in *config.pro*

Notice the icons in the first column beside the option names. These mean the following:

⚡ (lightning) - option takes effect immediately

✦ (wand) - option will take effect for the next object created

▣ (screen) - option will take effect the next time Pro/E is started

If you are using a *config* file from a previous version of Pro/E you may see a "stop sign" (actually a red circle with a line through it), which means that the option is no longer used.

Try to add an illegal option name. For example, in Release 2000i there was an option **sketcher_readme_alert**. Type that in to the **Option** field. When you try to set a value for this, it will not be accepted (the *Add/Change* button stays gray). Pro/E only recognizes valid option names! Thus, if you mistype or enter an invalid name, this is indicated by not being able to enter a value for it.

Saving Your *config.pro* Settings

To store the settings we have just created, select the *Save As* button 💾 at the top of the **Preferences** window. At the bottom of the new window, type in the desired name for the file - in this case *config.pro* and select *OK*.

Deleting Configuration Options

With the configuration file name visible in the Showing field at the top, highlight one of the options and select *Delete*. Selecting *Apply* automatically saves the new settings. *Close* the window.

Loading a Configuration File

To load a new configuration file, select the *Open File* button beside the **Showing** list. Select the desired file and then *Open*. Note that these settings will be read in but not activated immediately (note the green star). Select the *Apply* button and observe the green star.

Now select *Close* in the **Options** window.

Checking Your Configuration Options

Because some settings will not activate until Pro/E is restarted, many users will exit Pro/E after making changes to their *config.pro* file and then restart, just to make sure the settings are doing what they are supposed to. Do that now. This is not quite so critical since the window shows you with the lightning/wand/screen icons whether an option is active. However be aware of where Pro/E will look for the *config.pro* file on start-up, as discussed above. If you have saved *config.pro* in another working directory than the one you normally start in, then move it before starting Pro/E. On the other hand, if you have settings that you only want active when you are in a certain directory, keep a copy of *config.pro* there and load it once Pro/E has started up and you have changed to the desired directory. To keep things simple, and until you have plenty of experience with changing the configuration settings, it is usually better to have only one copy of *config.pro* in your startup directory.

Note that it is probably easier to make some changes to the environment for a single session using *Tools > Environment*. Also, as is often the case when learning to use new computer tools, don't try anything too adventurous with *config.pro* in the middle of a part or assembly creation session - you never know when an unanticipated effect might clobber your work!

Customizing the Interface

In addition to the environment settings, there are several ways of customizing the Pro/E interface: using *config.pro*, toolbars, menus, and mapkeys. An example of a customized interface is shown in the figure at the right. When you modify the interface layout, your changes will be saved in a *config.win* file in a directory of your choice (usually the current working directory). It is possible and permissible to have several different *config.win* files in different directories, each with a different customization of the screen to suit the work you may be doing on files in that directory.

In this section, we will introduce methods to customize the toolbars and menus.

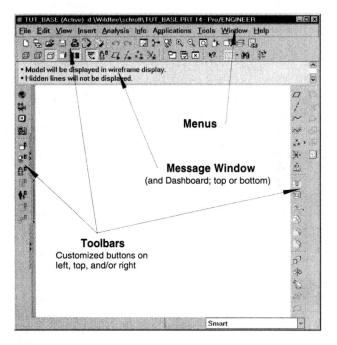

Figure 4 A (somewhat cluttered) customized screen layout

Toolbars

With the cursor on the top toolbar, hold down the right mouse button. This brings up the menu shown in Figure 5. This shows the toolbar groups currently displayed (see check marks); the groups can be toggled to include/exclude them from the display. Each group contains a set of functionally-related shortcut buttons.

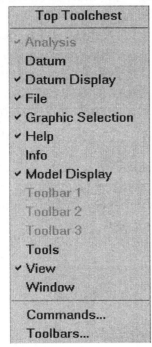

Figure 5 The **Toolbar** toggle menu

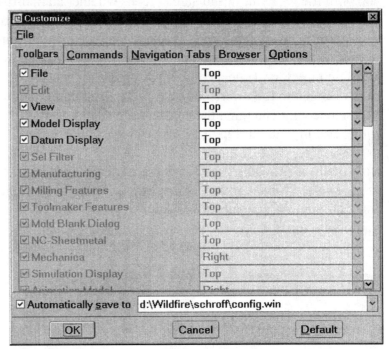

Figure 6 The **Toolbars** tab in the **Customize** window

At the bottom of this pop-up menu, select **Toolbars**. This brings up the **Customize** menu which contains a list of all available toolbars, and their location (see Figure 6). At the bottom of this window you can specify whether or not, and where, to automatically save the current layout settings. The default is *config.win* in the current working directory. As mentioned above, you can create multiple *config.win* files, and use **File > Save Settings** and **File > Open Settings** in the **Customize** window to store and recall previous files. Note that in addition to the eleven standard toolbar groups there are three initially empty groups (Toolbars 1 through 3), which you can populate with short-cut buttons using methods described below. The pull-down lists at the right allow you to place the selected toolbars at different places on the screen (left, right, top of graphics window).

Changing Toolbar Buttons

In the **Customize** window, select the **Commands** tab. (This is also available by selecting **Commands...** in the menu shown in Figure 5 or using **Tools > Customize Screen** in the pull-down menu.) The window shown in Figure 7 will open. Groups of toolbar commands are listed in a tree structure in the **Categories** area on the left. Click on any of the group names and the available short-cut buttons will appear in the **Commands** area on the right. As you move the mouse over these buttons, a tool tip will display.

To add a button to a toolbar, just drag and drop it onto an existing toolbar at the top, right, or left. The button will be added wherever you drop it on the toolbar. To remove it, drag it off the toolbar and drop it somewhere else (on the graphics window, for example). Note that it is possible to mix and match the short-cut buttons: any button can be placed on any toolbar. For example, a button listed under the **File** category can also be added to the **View** toolbar. Buttons can also be present on more than one toolbar. The possibilities are endless!

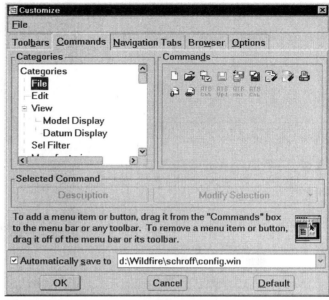

Figure 7 Choosing short-cut buttons to add to toolbars

At the bottom of the **Categories** list is **New Menu**. You can drag this up to the menu area at the top of the screen to create your own pull-down menus.

If you turn on one of the user toolbars (select Toolbar 1, 2, or 3 under the *Toolbars* tab), an initially empty button will appear in the designated location (top, left, or right). You can use the **Commands** selector to drag any button to define your own toolbar.

Check out the *Navigation* and *Browser* tabs. Notice that the final tab in the **Customize** window is *Options*. This lets you set the position of the dashboard and command/message window (above or below the graphics area) and some other settings.

When you leave the **Customize** dialog box, your new settings can be written to the file designated in the bottom text entry box. Each new or altered *config.win* file is numbered sequentially (*config.win.2, config.win.3*, and so on).

Helpful Hint

It is tempting, especially if you are blessed with a lot of screen space, to over-populate the toolbars by trying to arrange every commonly used command on the screen at once. This is reminiscent of many other Windows-based CAD programs. Before you do that, you should work with Pro/E for a while. You will find that Pro/E will generally bring up the appropriate toolbars for your current program status automatically. For example, if you are in Sketcher, the Sketcher short-cut buttons will appear. Thus, adding these buttons permanently to any toolbar is unnecessary and the buttons will be grayed out when you are not in Sketcher anyway - you are introducing screen clutter with no benefit. Furthermore, many commands are readily available in the right-mouse pop-up menus.

This concludes our short introduction to configuration files and customized toolbars. Hopefully, this has given you enough information to explore customization of the Pro/E interface on your own.

Index